Being Somewhere

Ferdinand Pöhlmann

Being Somewhere

Egocentric Spatial Representation as Self-Representation

1682 J.B. METZLER

Ferdinand Pöhlmann
Stuttgart
Germany

This work has been accepted as a doctoral dissertation by the Faculty of Philosophy at Eberhard Karls Universität, Tübingen in the winter semester 2015/2016.

ISBN 978-3-658-18018-8 ISBN 978-3-658-18019-5 (eBook)
DOI 10.1007/978-3-658-18019-5

Library of Congress Control Number: 2017941516

J.B. Metzler

Printed on acid-free paper

This J.B. Metzler imprint is published by Springer Nature
The registered company is Springer Fachmedien Wiesbaden GmbH
The registered company address is: Abraham-Lincoln-Str. 46, 65189 Wiesbaden, Germany

For Ilona, Hannah, and Jonathan

Contents

Preface

In the third part of this dissertation, I discuss the functioning of a certain kind of device and its significance for the claim that the ability to act is a constitutive condition on the capability for spatial perception: Sensory substitution devices allow blind persons to find their way with the help of sounds or tactual sensations. In the course of working on this thesis, I was lucky to be able to discuss my ideas with many colleagues who helped me not to get lost in the various mazes I had to cope with – similar to the way in which sensory substitution devices help the blind to orient.

In particular, I was fortunate to be part of a Research Training Group, funded by the Werner Reichardt Centre for Integrative Neuroscience (CIN), Tübingen where I collaborated with cognitive neuroscientists Hanspeter Mallot and Wolfgang Röhrich. They helped me to get an understanding of the brain mechanisms underlying spatial representations and the way in which science tackles issues in this area. Joining the Philosophy of Neuroscience Junior Research Group at the CIN offered me the opportunity to discuss my work with Chiara Brozzo, Krisztina Orbán, and Katia Samoilova, who gave me valuable comments on various occasions. Alex Morgan did a lot to help me structure my writing, especially the part on empirical evidence, and he always pushed me to clarify my thoughts and put things more clearly. Gregor Hochstetter not only has been a constant source of inspiration and constructive criticism; I am glad I can say that he has become a friend.

I was very lucky to have two extraordinary supervisors. I thank Frank Hofmann for inviting me to join the Research Training Group mentioned above, for helping me to develop the core questions of my thesis and to get an overview and a basic understanding of the pitfalls in the complicated discussion of self-consciousness and self-representation. To the same extent, I thank Hong Yu Wong for becoming my first supervisor after Frank had moved to Luxembourg, for his constant encouragement, criticism and faith in me, and for inviting me to become a member of his research group. This offered me the opportunity to organise and attend workshops and conferences with philosophers and scientists from all over the world, discuss my work with the excellent members of the group, and provided for a financial basis. Furthermore, to have Thomas Sattig as a third reviewer of my thesis has been an honour and I want to thank him for his carefully prepared report.

Beyond these people, who helped me to develop the topics of my thesis, I want to express my thanks to those who let me be able to concentrate on my work and went with me through the harder times. My parents supported me all along and made sure I never had to worry about my income. My most profound thanks, however, I owe to my wife Ilona – it was a long and sometimes challenging journey, I am very glad that she has been by my side.

1 Self-Representation

1.1 Introduction

The claim that shall be defended in this dissertation is that egocentric representation is to be understood as a kind of self-representation. As it stands, it is not obvious what significance this claim has in any way. To see that, let me sketch the broader picture that this enterprise is a part of.

The broad picture concerns the explanation of self-consciousness. One way to illuminate the peculiar characteristics of this phenomenon is to treat it as a special form of reference to oneself.[1] One important question that is debated among philosophers in this camp is what the peculiarity of the proposed self-reference essentially consists in. Here, notions such as immunity to error through misidentification, guaranteed right reference or the token-reflexive rule of 'I' play an important role. Another question that is debated here asks for the 'roots' of self-consciousness, or, as one might put it, the conditions of its possibility. It is to this debate that the present investigation shall contribute.

One influential approach towards an answer to this question regards self-consciousness as originating in social cognition, broadly speaking. The core idea here is that being conscious of oneself *qua* oneself is conditioned on being a member in a social community and being able to see oneself through the eyes of others, as it were.[2] An alternative to this proposal is to regard the

1 This is the currently predominant way, though it is not the only one. An alternative can be found in works that stand in a more German idealist tradition, e.g. Fichte (1970), Nozick (1981), Richards (1984), Lang (2010). According to these views, self-consciousness is to be explained in terms of a spontaneous act of the subject.

2 This approach is notoriously present in the philosophy of George Herbert Mead. It is also pursued by Musholt (2012), Lailach-Hennrich (2013), Rochat (2009), Beckermann (2003) and Neisser (1995). Tugendhat (1979) has been interpreted in this way, although the author rejected this interpretation recently (Tugendhat 2005).

roots of self-awareness as lying in spatial cognition.[3] The basic assumption that grounds the present investigation is that this latter strategy is most promising. I try to argue what may be regarded as one part of an account that posits spatial cognition as the basis of self-consciousness, although it is not possible to argue for the claim in its entirety here.

There are two senses in which the present investigation falls short of being a comprehensive argument to the effect that the roots of self-consciousness have to be seen as lying in the domain of spatial cognition. The first pertains to the forms of spatial representation that are subject of this dissertation. Generally, it is common to differentiate between two basic kinds of spatial representation, namely egocentric and allocentric. I will confine myself here to the first kind. Egocentric spatial representation has two essential features. The first consists in that the represented spatial information is indexed with respect to the representing subject, i.e. that spatial content of the representation is sensitive to the spatial relation the subject bears to what it represents. The second essential feature is the immediate relevance for behaviour that these representations have. It is not only the case that descriptions of what one does are naturally expressed in part by egocentric terms, such as 'left', 'right', 'in front', 'up' and so on. Moreover, representations of spatial relations that cannot be 'translated' into egocentric representations cannot be used to guide the behaviour of the representing subject (at least when we take this behaviour as consisting in acts of the subject that are directed at the represented object). One has to know where something is in relation to oneself to act on it.

I take it that egocentric spatial representation is not the only possible starting point for such a project. One could also argue from some representation of one's own place as part of a representation of the larger environment one inhabits, something that is associated with the notion of a cognitive map. However, starting with egocentric space is in a way more fundamental since the capability to represent space egocentrically is commonly regarded as more basic (and, therefore, more widespread among the animal kingdom) than the

3 This idea is inspired by Kant's philosophy and present in Neo-Kantian works such as Strawson (1959), Evans (1982), or Cassam (1999).

capability to represent space in ways that transcend the actual perspective of the subject.

The second sense in which the present investigation will fall short of presenting a comprehensive explication of spatial representation as the root of self-awareness is the following. The notion of self-representation, for which I argue in this dissertation that egocentric representation is a subclass of, is more general than the notion of self-conscious representation. However, what will not be a proper part of this dissertation is an account of how exactly self-conscious representations, i.e. the kind instances of which are expressible with the first-person pronoun, is related to the general kind of self-representation. The general background assumption here is that the explication of self-representation as a not-yet-self-conscious representation is an important step towards elucidating some important features of the way self-reference functions in self-conscious representations.[4]

The principal aim will be to explicate egocentric spatial representation as a kind of self-representation. The first step will be to explore the notion of self-representation, which I will take to be situated between mere *de re* representations and *de se* representations. The basic idea is that self-representations are those that are treated as pertaining to the subject itself (without this necessarily being understood, i.e. without there necessarily being a representation of oneself *as* oneself).

I will then turn to egocentric spatial representation and will present three different lines of thought that may result in taking it as a subclass of self-representation. In the course of discussing these arguments it will be clarified what is peculiar of egocentric representation, how its content has to be specified and what its constitutive conditions are. I will first discuss an argument that is grounded in positing relational spatial properties as the content of egocentric representation. Since in this case the subject is explicitly represented, it seems natural to regard egocentric representation as self-representation on this proposal. However, there are compelling reasons against this relationist

4 For a recent attempt to explain the connections between basic forms of self-representation and mature self-consciousness, see Peacocke (2014).

picture and in favour of taking the spatial properties that are represented in egocentric representation as monadic. While being more satisfying as an analysis of egocentric content, this approach renders it dubious whether one can still take egocentric representation as self-representation. The reason is that there is no element in the content anymore that refers to the representing subject.

In what follows I will discuss two arguments to the end that egocentric representation can indeed be regarded as self-representational, although there is no singular element in the content that refers to the representing subject. The basic strategy of both these arguments is to look not on the content but on the constitutive conditions of egocentric representation.

Several thinkers proposed a line of thought that might be regarded as an 'argument from design'. It takes as premise the peculiarities of the information processing machinery that grounds egocentric representation and argues that egocentric representation counts as self-representation since it only ever concerns the representing subject, even without there being a singular element in the content that explicitly refers to the subject. As attractive as this position might seem, however, I will argue against it on the grounds that it doesn't get the constitutive conditions on egocentric representation right.

Beginning with Part 2, I will argue that a sense of one's own capabilities to move is a constitutive condition on egocentric spatial representation. This is the second basic claim that I want to defend in this dissertation. Most chapters are devoted to it. To arrive at this claim I will present both a historic and systematic overview of approaches that argue that (embodied) action has to be taken as a constitutive condition on egocentric representation of space and that a mere explication of the information processing machinery is not sufficient. As far as I know this is the first comprehensive presentation that contrasts these views. In particular, I will discuss the approaches taken by Edmund Husserl, Max Scheler, Maurice Merleau-Ponty, James J. Gibson, Gareth Evans, Susan Hurley, Alva Noë, and Shaun Gallagher. The accounts will be groped according to which feature of action is taken to be decisive.

In Part 3, I will evaluate empirical evidence that speaks in favour of this approach as well as empirical findings that support a contrary view. The evidence includes, among others, studies on inverted vision and sensory substitution devices. My conclusion will be that there are strong reasons to regard the capability to act as a constitutive condition on egocentric representation.

In the final Part 4 of my dissertation, I will develop the details of my position and argue for the claim that it is not the bare capability but a *sense* of one's own capabilities to move that partly constitutes egocentric spatial representation. To this end, I will inter alia draw on results from developmental psychology. The last step of the dissertation will be to show that it is a natural consequence of this view about the constitutive conditions of egocentric spatial representation that it is a kind of self-representation. Since self-representation will be defined by their immediate significance for action, nothing could be better suited to play that role than a kind of representation whose instances are already shaped by the subject's sense of her own capabilities. Thus, this last step establishes the self-representational nature of egocentric spatial representation. It presents one first and crucial step to an explanation of the origins of self-consciousness as lying in egocentric representation of space.

1.2 Egocentric spatial representation as self-representation I: a relationist view

In this chapter and the next one, I want to consider two ways in which one might conceive of egocentric representation as self-representation. In the present chapter, I will begin with an argument that is based on a relationist understanding of egocentric spatial representation. According to this argument, egocentric representation is self-representational since it explicitly refers to the representing subject. Discussing this argument will lead to a clarification of the notions of egocentric representation and self-representation, respectively. In the subsequent chapter, I will consider a second argument for the self-representational character of egocentric representation. This one will be based on the constitutive conditions of egocentric representation – not on its content. It argues that the design of the cognitive machinery that subserves egocentric representations implies the self-representational nature of egocentric representation. Discussing this argument will help to better understand the overall approach that I want to take in this dissertation since my own argument also rests on an understanding of the constitutive conditions on egocentric representation.

I will criticise both arguments, however. I will argue that the first is based on a misunderstanding of the contents of egocentric representations, and the second gets the constitutive conditions of egocentric representation wrong. Thus in one sense, both of these two chapters develop negative argumentations; there is no complete positive argument for the self-representational nature of egocentric representation given yet. However, as already said, there are important positive aspects included in these chapters as well. Important basic notions will be clarified and some parts of the argument in the next chapter I will keep for my own argument. The remainder of the dissertation then, beginning with Ch. 2.1, will explore a superior way of arguing for the self-representational character of egocentric representation that is based on taking action as a constitutive condition on egocentric representation.

What do we mean when we speak about egocentric spatial representation? As already mentioned in the introduction, an intuitive understanding is at hand, which states that egocentric representations of space are those that represent spatial properties in relation to the subject, thus underpinning the distinctive perspective that every representing organism has of its surroundings and enabling instant action control towards represented objects. This intuitive idea of what egocentric representation amounts to directly leads to the first proposal of how to analyse the content of these representations, namely in relational terms. On this relationist account, representing something as being towards the right means to represent the content 'that x is to the right of S', where 'S' refers to the subject itself and 'x' refers to the represented object. This way of spelling out the content of the representation includes an explicit reference to the representing subject herself. Given such an analysis, it is not difficult to see how egocentric representation can be understood as an instance of self-representation: since it explicitly refers to the subject, the case would seem straightforward.

However, I want to argue against this approach on the basis that the contents of egocentric representation are not to be taken as relational, but as monadic. On this approach, representing something as being to the right simply means to represent the content 'that x is to the right'. This way of spelling out the content does not include any explicit reference to the representing subject, however. Therefore, the question arises of how to explain that these representations can nevertheless count as instances of self-representation.

To answer this question I will turn toward explicating the notion of self-representation. 'Self-representation' intuitively comprises representations that play an important and distinctive role in the lives of cognitive organisms. They are linked to actions of these organisms in special ways. Representing something as pertaining to oneself arguably has more immediate implications for action than representing something as pertaining to someone or something other. For example, to represent a car as coming towards oneself immediately leads to certain actions in a way representing a car approaching someone else does not. Similarly, to represent one's own pants as being on fire elicits different actions than representing that someone else's pants are on fire. The most salient instance of self-representation is one we would verbally express with the first person concept 'I'. However, there are other representations that plausibly fall under the label of self-representation that are not that sophisticated. Representing own bodily states, or representing the place one currently is at, are examples for such representations that we would consider to be self-representations because of their immediate significance for the subject's actions. These latter kinds might be analysed without the first person and are therefore applicable to lower cognitive states or systems. Thus, arguably not any self-representation is *per se* a self-conscious representation. One might suspect that the kind self-representation comprises those representations that explicitly refer to the representing system itself, i.e. whose content includes some expression that stands for the representing system. On this view, all representations that explicitly refer to the representing system itself would count as self-representation. However, based on considerations of John Perry and Ernst Mach, I want to show that this approach cannot be appropriate. It does not fit the assumed peculiar functional roles that self-representations have. Therefore, both of these initially plausible ways of cashing out the notion of self-representation, with the help of 'I' or merely as a reflexive representation, are not appropriate. The first is implausibly sophisticated while the second is implausibly simple. Another way of characterising self-representation is in order.

What I want to explicate is that it is not the content that is distinctive of self-representation but the special functional roles that these representations play in the cognitive lives of organisms. I will first consider both of the mentioned proposals in more detail and explain why they are not appropriate. Thus, in section 1.2.3 I will begin to defend the view that there is a way to cash out the notion of self-representation that does not allude to the first person but preserves the special status of self-representation. The notion I want to defend is situated between mere reflexive representations and full-blown self-conscious representations.

This will be the somewhat open end of this chapter. On the one hand, the monadic approach toward explaining egocentric representation seems more appropriate as an account of how spatial properties are represented in egocentric spatial representation than the relational view. On the other hand, given this approach, it is obscure how it is possible to think of egocentric representation as self-representation, since on this account there is no explicit representation of the representing subject itself as there is on the relational account. Although I will have presented some reasons for regarding self-representation as not depending on any explicit reference to the representing subject then, the details of the positive view of how to understand egocentric representation as self-representation will be presented only in the next chapter.

In Ch. 1.3, I will discuss one way of explicating egocentric representation as self-representation that is based on the monadic analysis of the content of egocentric representation and is based on the peculiar design of the cognitive machinery that subserves egocentric representation.

1.2.1 Egocentric representation: A relationist view

The first attempt of explicating egocentric spatial representation as self-representational is based on the idea that these representations have relational contents. To understand this proposal, I will first explicate the notion of egocentric space more thoroughly in this section. The next step is to consider the relational nature of the contents of egocentric representations. On this account, egocentric spatial representations explicitly refer to the representing subject. This explicit reference might be seen as a good reason for regarding egocentric representation as self-representation. In the second section of this chapter, however, I will argue that the contents are plausibly *not* to be taken as relational but as monadic. This undermines this first attempt of arguing for the self-representational character of egocentric representation based on the supposed nature of their contents.

1.2.1.1 Egocentric representation: initial considerations

Representations of space that specify the spatial relation between an object and the observer are commonly termed 'egocentric representation'. The idea is that the representing system forms a point of reference in respect to which distances and directions to perceived objects are specified. In this sense, egocentric spatial representations are perspectival, that is, their content essentially depends on the location of the representing subject in relation to the perceived objects and it changes in accordance with changes in this relation. The contents of egocentric spatial representations, that is, the conditions under which any such representation is to be regarded as correct, depend on where in the world the subject is and how it is oriented.

Not all spatial representations are of this sort. Spatial relationships of objects can be represented in a different way, a way that is not tied to the location and orientation of the representing subject. These representations specify distances and directions between objects based on another frame of reference. What is needed at least is a reference direction and some measure for distance. With these means, one can specify the position of an object x with respect to an object z by specifying its distance and direction, the latter in terms of deviation from the reference direction. The represented spatial relations do not change in case the representing subject changes its location, *ceteris paribus*. That means that in case x and z are regarded as stable objects or locations such as two buildings or natural places, like hills or trees, the representation of their spatial relationship is stable. Egocentric representations of space, in contrast, change whenever the subject moves.

The difference between these two kinds of representing space comes out most clearly by considering their respective significance for action. Suppose one represents some object x egocentrically as being located in such and such direction at such and such distance and one furthermore knows that there is an object z at such and such location specified in relation to x. Suppose you are in an amusement park. You came a little late there but you and your friends agreed on the phone to meet at the chip shop. You are there for the first time. Your friends told you that the chip shop is about fifty metres left of the great wheel. You enter the park, it is crowded and noise comes from every direction from any of the attractions around you. You do not know where to go. You look around through all these happy people with their even happier kids until you finally spot the big wheel beyond these masses of people. Now you know where to go. There is the big wheel and fifty metres left must be the chip shop; nothing easier than this.

What is apparent from this scenario is that you are only able to act in a reasonable way when you see the big wheel. (Of course, you could also have simply started walking somewhere and try to find the chip shop in a trial-and-error fashion. I leave that irrational option aside.) Of the two relevant spatial representations you had at the time, the one specifying the relation between the chip shop and the big wheel and the other one specifying the relation between you and the big wheel, only the latter is poised to guide your action. Either you would head on to the big wheel and then go left, or, if it turned out there is another way, you could also make up a more direct route that ends fifty metres left of the big wheel by means of a simple triangulation that relies on your representation of your location relative to the big wheel and the representation of the spatial relation between the chip shop and the big wheel. However, none of these possibilities is open if you do not know where the big wheel is relative to you (or, where something else is of which you know the egocentric location and that objects' location relative to the location of the big wheel). To act on an object one has to represent it egocentrically. To act on an object that one currently does not represent egocentrically one has to act on another object, or at least, indirectly, based on another object, that one represents egocentrically. There is always a 'gap' between representing some spatial relation allocentrically and being able to act on it. You always have to know the egocentric relation to be able to act. Thus, egocentric representation closes the gap between spatial representation and action. Hence, the important distinction between egocentric and non-egocentric spatial representation lies in the immediate relevance for action that the one has and the other lacks (Campbell 1994, Ch. 1; Campbell 1993).

1.2.1.2 Relational content

One might suppose that the content of egocentric spatial representations has to be specified in the following, relational way. 'S represents x egocentrically' would mean that the subject e.g. represents "that x is about 100 metres straight ahead and slightly to the left of me"[5]. This would not only seem intuitively plausible but furthermore it would obviously present a good candidate for an instance of self-representation. The representation is not only appropriate to directly guide the actions of the subject if it was his intention to approach x, for example, but it also explicitly refers to the representing system. Since the expression 'me' is part of the content, it is even plausible to regard this representation as a self-conscious one.

The question I want to pose first is whether it is necessary for this kind of representation, namely egocentric spatial representation, to include some form of the first-person concept. That is, is it necessary that egocentric spatial representation is only possible for creatures capable of self-consciousness?

This view seems to me to be highly counter-intuitive. The capability to represent egocentric spatial information is commonly regarded to be the most basic and primitive form of spatial representation (Pick Jr 1993; Mallot and Basten 2009). It seems, whenever we can attribute to some organism representational states that function to represent the outer world we are primarily inclined to attribute states that represent spatial relations between the subject and the objects it encounters (Burge 2010, Ch. 9). One might argue, e.g., that representational states are at least in play when an organism is able to keep track of the position of some object which it has encountered before but which is now 'out of sight'. In the attempt to attack some prey, it might be necessary for the predator to search for some location from which it can attack best. This search in many instances might include moving in a way around the prey in which it is impossible to perceive the prey continuously. To accomplish its plan, the predator as it moves needs to keep track of the prey's location. These cases are straightforwardly explained by supposing

5 I do not want to claim that distances are represented in that mathematical way in egocentric representations necessarily, however.

some internal state on the side of the predator that represents the location of the prey and changes according to the movements of the predator.[6] It seems very likely that capabilities such as these are far more widespread in animal kingdom than the capacity for self-conscious representation.

A better way to describe the structure of egocentric representation would be to retain the above given structure but replace the occurrence of the first-person concept with some other expression that refers to the subject, while leaving it open whether it is a self-conscious representation or not. Thus, 'S represents x egocentrically' would mean that the subject represents "that x is about 100 metres straight ahead and slightly to the left of S". The substitution of 'me' through 'S' in the specification of the content clearly renders this representation as ambiguous between being self-conscious and merely being reflexive. However, it has been argued that this way of stating the content is already expressive of a self-conscious state (Anderson and Perlis 2005; Beckermann 2003).[7] I do not think that this is correct. The natural way to classify representations like this would be to say that they are instances of a *de re* representation of the representing subject. What is needed to explain some action as the one of the predator above is just a representation of the spatial relation between the subject and its goal object. Why should one demand of such a representation that it includes some self-conscious form of self-reference? This seems unnecessarily sophisticated as I already said. Furthermore, the performance of the act that is supposed to be based on the egocentric representation would not benefit in case the representation is a self-conscious one. What more relevant information for the task at hand is part of the supposed self-conscious representation in contrast to the mere self-representational representation? With respect to their action-guiding potential (at least with regard to bodily actions towards represented objects in one's vicinity), there

6 A somewhat similar example is given by Burge (2010), pp. 515–7, who explicates the behaviour of a jumping spider (Portia). It can be used to illustrate the case, although Burge stresses the generation and exploitation of representations of routes.

7 For a criticism of Beckermann's view (and the whole approach to explain self-consciousness via social cognition) cf. Hofmann and Pöhlmann (2013).

seems to be no difference at all. Thus, one should be reluctant to think of these states as self-conscious.

This is not to say that this way of spelling out the content might not be expressive of a self-conscious state. It can. However, as it stands, it is at least neutral regarding the question of self-consciousness. When the content of some representation has to be spelled out with the help of some form of the first person, this representation can be taken as expressive of self-consciousness. Where the content is merely spelled out with the help of one element that refers to the representing subject, the question cannot be answered that straightforwardly.

Nevertheless, in specifying the content that way, even without supposing that it implies self-consciousness, one has good reasons to regard egocentric representations in general as instances of self-representation. Slightly more formally put, the following would be the case. Egocentric spatial representation is a kind of spatial representation that specifies the spatial relation (i.e. distance, direction) between subject and represented object. The content of this kind of representation is generally to be specified as follows: "S represents x egocentrically", where x is some everyday object, means that "S represents that x is located in such-and-such direction and in such-and-such distance from S". Two features that seem initially to be essential for self-representation are clearly in place. First, the representation explicitly refers to the representing system itself, as is clear from the appearance of 'S' in the content. Second, it also is poised to be used to directly guide movements by the subject, since it contains spatial information that is specified relative to the representing subject and not to some other object in the world.

With respect to the basic question of this section about the role of self-representation in spatial cognition, this would mean an answer that is as clear as simple. On this relationist proposal, egocentric spatial representation is obviously just a subclass of self-representation. Every instance of egocentric spatial representation includes an explicit representation of the representing subject itself and is therefore an instance of self-representation.

1.2.2 A challenge: monadic spatial contents

As nice and straightforward as this answer to the question would be, it seems to me that the story is not that simple after all. The reason is the following. There are independent grounds for supposing that the content of egocentric spatial representation has to be specified non-relationally. Naturally, this does not mean that the object that is represented egocentrically in this case does not stand in a spatial relation to the subject. The basic idea behind this proposal is that what is represented is indeed a spatial relation between the subject and a given object but that this is not explicitly reflected in the content of the representation. The issue is not an ontological but a representational one. It is not about spatial relations but the representation of spatial relations.

The proposal could be expressed by making the following claim. Egocentric spatial representations do represent objects around the perceiver in a non-relational way. That means that objects are represented as "to the left" or "to the right" or "in front and slightly left" etc. The crucial difference to the relational proposal above is the missing explicit *relatum* in the content, that earlier was 'me' or 'S', respectively.

What reasons are given for this proposal? They are not explicitly stated anywhere by those who endorse the view.[8] So one has to read between the lines to a certain extent to figure out their reasons. I think there are largely phenomenological points, besides considerations about ecological and developmental plausibility and recourses to simplicity.

1.2.2.1 Simplicity

Let me start with the last one. Perry and Beckermann frequently say that there is no "need" to explicitly represent the subject herself in the case of egocentric representation (Beckermann 2003, p. 179; Perry 1986, pp. 150–51; 1998b, p. 87). This way of analysing these cases is often contrasted in their writings with ways where the representation explicitly refers to the subject, and even

8 The goal of one of the central papers of Perry's I rely on here merely is to explore the possibility of the monadic view: Perry (1986), cf. esp. p. 138; Beckermann (2003). The approach is also present but merely stated in Campbell (1994), p. 119; cf. Campbell (2002), p. 47. For discussion, cf. Dokic (2015).

with the concept of 'I'. In this context, the motivation seems to be something like Ockham's razor. To give a theory that has the same explanatory force than another one while at the same time being simpler is to be regarded as superior. Since it is possible to explain the features of egocentric representation without supposing any explicit reference, which would make thinks more complicated, the monadic approach is to be taken.

However, contrary to first impression, it seems doubtful whether this is their real motivation and whether it would be a particularly good one at all. In one sense of simplicity there is no difference between both approaches, namely in the number of postulated entities. The relational analysis does not posit some additional entity, like a self, in their explanation, which the monadic view does not. Rather, the approaches only diverge in terms of the complexity of their analysis of the relevant contents. It seems to me that basing the decision only on this rather subtle difference and inferring from there that the monadic approach is better would be a very weak option. It would be much better to have something substantive to say why one's own approach is superior. I think that on a more sympathetic reading of their writings one can make out these rather substantive reasons.

1.2.2.2 Ecological and developmental plausibility

The talk about no need of explicit reference can be understood in a more substantial way, too. It is frequently accompanied by the idea that there is accordingly no need to keep track of the subject (Perry 1998b, p. 87; Beckermann 2003, p. 179). It suffices, on this account, that the subject keeps track of the represented object. In this perspective, the claim is not about the theory in general but it arguably shall raise the ecological or developmental plausibility of the approach, as one might put it. What stands behind this point is the plausible conviction that animals or children enjoy egocentric representations as we do. An explanation of this that gets along without positing explicit reference to the subject in these cases can be viewed as more plausible since it is supposedly less cognitively demanding. This is the point of talking about "no need to keep track," I take it. Keeping track of objects is a demanding task, thus, an explanation of egocentric representations that does posit less complicated tasks of keeping track seems more appropriate to explain these rather primitive cases than alternative approaches that imply higher efforts of keeping track.

Now it seems to me that plausibility claims about cognitive demand are rather difficult to assess without an appropriate measure of it and without any criterion for when cognitive demand has to be regarded as too high. Neither is given by Perry or Beckermann. This makes their claims quite speculative. It would first have to be shown that explicitly representing oneself, even in a non-self-conscious way, is such a demanding task that it cannot be assumed to be possible for more primitive cognitive systems. Specifically Beckermann, who gives a quasi-developmental story, simply assumes that this is intuitively evident. But isn't it equally intuitively plausible to assume that the most basic representations are explicitly about oneself precisely for the reason that they are based on ways of processing information that necessarily concern the subject itself? This is idea is part of Anderson and Perlis's account, one that we will encounter below, and they too seem to allude to intuitive plausibility.

I think to settle the question whether a selfless account is more promising or plausible along these lines one would have to come up with some considerations and empirical evidence about the demands that are necessary to represent oneself explicitly. These could be given in two ways as far as I see. Either, one could start with the premise that representing oneself just is as demanding as other instances of representing objects. Then there would merely be a quantitative difference between merely representing objects at locations and representing objects at location relative to oneself. One would have to employ one skill more often in the latter case than in the former. I think, however, that Beckermann wants to draw a stronger contrast here.

Arguably, he wants to say that explicitly representing oneself embodies a different ability that the subject has. On Beckermann's account, what is needed to explicitly represent oneself is a representation of others' cognitive states about oneself. Thus, the subject has to realize that there are other *subjects* inhabiting his world that represent him in the way he represents them. He has to have metacognitive abilities in the sense that he represents others' cognitive states and not mere objects and features of the environment he can perceive. This would mean that representing oneself (even in a non-self-conscious way) is different in kind from representing some random object in the vicinity. The difference would then be qualitative instead of merely quantitative. The cognitive system that is able to explicitly represent itself could be regarded as existing on a higher developed level than the one that is not able to represent itself explicitly. In this second way, the approach of Beckermann seems more promising and persuasive since it better fits the quasi-developmental story he wants to tell.[9] However, as it stands, it only gives some direction for further inquiry. The crucial question then becomes what it really means to be capable of self-representation and whether it really makes necessary some explicit reference, as Beckermann seems to assume. I understand my dissertation partly as a contribution to that question.

Overall, I think that this second way of arguing against the need of explicit self-representation has much more theoretical potential than the first one that is merely based on considerations of theoretical simplicity. Here, the reason is more substantive in that it's supposed to make intelligible primitive cases of egocentric representation that are intuitively plausible. However, as I said before, to be really convincing the account would have to include some justified claims about the demands of explicit self-representation. Just to regard these self-less representations as a mere theoretical possibility that is intuitively plausible and not impossible is not enough to show that primitive egocentric representation really is to be analysed in this way.

Having said these rather critical points, I want to present the next, phenomenological considerations that pull in a positive direction and also help to render these other claims somewhat more plausible.

9 I take it that he does not really want to give a developmental account, neither phylo- nor ontogenetic. It is more like an exploration of plausible possibilities and a conceptual endeavour to find the least demanding form of representation that can be said to be about the subject. In that sense, it is an exploration of constitutive conditions for egocentric representation.

1.2.2.3 Phenomenology

I think the most promising reason for the monadic view is based on phenomenological observations. Perry writes that "a natural way for me to report what I saw [an apple before me; F.P.] would be to simply say: 'That's an apple.' or: 'There is an apple there'" (Perry 1998b, p. 86). This observation is also made by John Campbell. He states that the "egocentric frame used in vision employs monadic spatial notions, such as 'to the right', 'to the left', 'above', 'in front', and so on, rather than relational notions, such as 'to my right', 'above me', 'in front of me', and so on" (Campbell 1994, p. 119; cf. Casullo 1986; Schellenberg 2007, p. 621; Cassam 1999, p. 78; Dokic 2015). The claim here is that the appropriate description of my visual experience does not include any reference to me, or the representing system, respectively. That is supposed to mean not only that it is possible that egocentric contents do not explicitly refer to oneself but that it is indeed highly plausible that they do not, given the phenomenology of perception. On this view, the most basic form of egocentric representation is to be regarded as 'selfless'. Any representation that explicitly refers to the representing subject is then to be regarded as more sophisticated or of a different kind, e.g. a belief. Let me consider this line of thought in more detail.

With the first few sentences of his "Thought without representation" (1986)[10] Perry describes an everyday situation of seeing a cup in front of oneself, reaching out, picking it up and drinking from it. He wonders how he can do so based on his representation of the cup, since apparently, he is not part of the visual field. His conviction, which first of all leads to his wonder of how he is able to pick up the cup, is that he is not present in what he sees. All one sees in such a situation is the cup, there on the table. It does not seem to be the case that one would describe the situation as 'seeing a cup in front of oneself'; at least this does not seem to be the simplest way to describe the situation. It is just 'there'. One might hold that Perry's description of the sit-

10 The same basic line of thought is also present in more recent articles by Perry, e.g. Perry (2011).

uation is the most natural to give. Therefore, one might argue from a phenomenological perspective that one is not part of the content of ordinary egocentric spatial representations.

Is the subject part of the visual field?
The case is not as straightforward, though. For one thing, opponents of Perry's view might argue that one *is* actually part of the visual field. One sees the borders of one's nose, for example (nicely reflected in the famous drawing of Ernst Mach's which shows him sitting on his couch with a pencil in his right hand as supposedly seen by himself and that so often has to serve as illustration for the phenomenology of perception, see Figure 1). In response, defenders of Perry's view could reply that the borders of the nose are phenomenologically transparent so that they ordinarily are not part of the content of visual perceptual representations. They are only present in cases where attention is focused on them.

Figure 1: Illustration of egocentric perspective (taken from Mach 1922, p. 15).

However, the opponents might respond that other parts of the body might be part of the content. In the example, Perry supposedly sees his own hand simultaneously with the cup. This could also explain very well how action is controlled by this representation. However, Perry might react in asking whether it is really the case that we assert that we see the cup in front of us because we see it in relation to some part of our body, e.g. our hand or even our nose. It seems not. Suppose it is completely dark, you only see some small light back there slightly to the left, nothing else. This representation is also naturally classified as egocentric and it can also guide you in moving towards the light. Moreover, this is presumably without any body part of yours being present in the visual field and therefore you are not being represented in the visual perceptual content. Therefore, it seems that there are cases where one is not part of the visual representational content but where the representation is plausibly classified as egocentric nevertheless. Visual egocentric spatial representation thus appears not to be dependent on there being some visual representation of the subject itself in it. One might generalize this finding in support of Perry's view and argue that even if parts of one's body happen to be part of the visual field and thus are supposedly also part of the visual perceptual content, they are so not in the relevant way, namely as providing the reference point for the egocentric frame of reference. Thus, the content of the representation is then to be expressed as "my hand in front and the cup slightly behind" but not as "the cup in front of my hand".

A similar line of thought might be embraced with respect to another sensory modality than vision. Auditory perceptual representations are also instances of egocentric spatial representation, although arguably somewhat diminished in that only the direction of the sound is specified whereas distance of the sound source can only be guessed or inferred based on prior knowledge. In hearing something, one usually perceives the sound as coming from a certain direction.[11] Here, it seems quite clear that the subject itself is not represented in the auditory field. One does not hear oneself and the sound simultaneously. Hence, as in at least some cases of vision, one is not part of the

11 Very low sounds are quite difficult if not impossible to locate, though.

sensory field. The content is intuitively articulated by something like "the sound comes from the left".

So far I have considered two objections against the monadic view that are both based on the conviction that the subject is somehow part of the visual field and *a fortiori* part of the content of visual egocentric representation. However, I do not think that these objections are sufficient to undermine the monadic view. Next, I will consider another possible objection that centres on a possible role of proprioceptive representations as underpinning the self-referential element in the content of egocentric representation. However, I think that this line of thought is mistaken, too.

Is the subject part of the broader perceptual content?

Proponents of the relational view might broaden the picture a bit and argue that although it is right that one is not in the right way part of the *visual* perceptual content in egocentric representation, one is part of a *more general* perceptual content that includes two explicit *relata*, namely the perceived object and the subject. In particular, the body, and especially the spatial relations between, and movements of limbs are represented in proprioceptive states. One might cast these as representations of the way the body is organized in space at a certain instant in time.[12] Opponents of Perry's view might then argue that these proprioceptive representations are integrated with perceptual representations of the outer world and become part of a general perceptual state, one that comprises any perceptual state in a given moment, and, therefore, the visual perceptual representation of the cup as well. Thus, the opponents would argue that there is a supramodal representational state, in which one's relation to the cup is explicitly (i.e., two *relata* featuring in the content) represented as that relation by means of combining proprioceptive and visual states. The content thus could be arguably expressed as "the cup in front of me", where the content is a combination of two distinctive representational elements with different aetiologies. The part which refers to the cup is grounded in a visual perceptual representation of the cup on the table, while

12 Cf. O'Shaughnessy's short-term body image, e.g. in O'Shaughnessy (1995).

the part that refers to the subject is grounded in a proprioceptive representation. Of course, the opponents have to tell some story about how the spatial parameters are integrated, since supposedly the spatial relationships among body parts are represented in a different frame of reference than the location of external objects. However, this might be very well possible and would pose no big problem for this approach, I think.[13]

To be successful, the opponents would have to argue that this is not just the case in this special situation, but generally (otherwise, Perry could simply come up with another example). Thus, they would have to argue that proprioceptive representations of the own body are part and parcel of any egocentric spatial representation and that therefore egocentric spatial content can and is to be regarded as explicitly relational, specifying spatial relations as e.g. "to the left of me". The view could be based upon accounts that endorse multisensory integration in explaining perceptual states (Ernst and Bülthoff 2004; de Vignemont 2014).

How could Perry respond to such a proposal? I think there are three responses he could have to such a line of thought. He could point toward imaginative states where the analysis seems implausible, cite pathological cases of patients that have lost their proprioception but retained egocentric representations, or argue that the argument changes the subject matter and therefore is invalid. I will consider these three points in somewhat more detail in turn.

1st response: imagination

First, he might argue in response that there are cases of egocentric spatial representation where there is no proprioceptive representation. Think of imaginative states, in which you imagine yourself standing somewhere, facing

13 For example, one might hold that the position of an external object is represented as primarily located with reference to the 'cyclopian point', which marks some point between the eyes that serves as the point of reference for distance and direction measures in visual perception. Since the spatial relationships of this point to other parts of the body are in turn part of the proprioceptive representations that specifies spatial relations between the different body parts as we said, it seems easily possible to integrate both spatial frameworks on such a picture.

some objects, moving past them etc. It seems that those could very well count as egocentric representations. However, is it plausible to assume that they represent spatial relations relative to oneself because proprioceptive representation is combined with imagined visual representation? Opponents might say that there is some imaginative proprioceptive representation in the same way there is imaginative visual representation here. Thus, one would represent oneself as an imagined embodied subject in these cases. Nevertheless, one might again ask from a phenomenological standpoint whether that reflects the phenomenology of imagination in all cases and if there are not cases where one just sees objects arranged in space without either imagining parts of oneself or imagining oneself as being embodied. It is not easy to decide here. At least it casts some doubt on the proposal that there always has to be some proprioceptive representation present in cases of egocentric spatial representation.

2nd response: deafferentation

I think, proponents of Perry's view can argue on a somewhat stronger basis against this proposal by considering pathological cases like the one of IW, the famous patient suffering from deafferentation (Cole and Paillard 1995). IW does not have proprioceptive representation of large parts of his body, that is, he does not know how his limbs are situated in relation to one another and he does not recognize if they are moving. (This would be somewhat parallel to the case above where one only sees the point of light in the dark and nothing else. There, one had no visual perception of the body; in this case, IW has no proprioceptive perception of the body.) Because of this, he constantly has to concentrate on what he does and how he moves, even in such simple tasks as walking. His strategy is to guide his actions solely based on vision. He has to constantly look at his legs when walking or at his hand when writing etc. because otherwise he would not know where he would move. However, it seems clear from the descriptions of his case that he is able to locate objects in the environment visually and to use such information to guide his actions, even though he has to visually represent both the target object and the effector, e.g. his hand, to be able to interact with the seen object. He is even able and allowed to drive a car. It seems that one can grant him an egocentric visual representation, despite his lack of proprioceptive representations. Therefore, at least for this case the content of the egocentric representation seems not to be specified as e.g. 'the cup in front of me' but only as 'the cup in front'. The representations that would be the basis of the 'me', i.e. the singular representation that refers to the subject itself, is simply missing; therefore, the 'me' cannot be part of the content.

One might object here that the case of IW does not present a case of total loss of proprioception and push the fact that he has some proprioceptive input from the neck up (Gallagher 2005, p. 63). Again, it is not clear whether highlighting this point is capable of securing the relationist's account. It is unclear to me whether it is intelligible that the highly diminished proprioceptive basis would be enough to ground the singular element in the content that refers to the subject. Is not the spatial information conveyed by egocentric representations usually indexed to the trunk of the subject? Think about how you have to change your position to perceive of some object that it changes its location relative to you. A rotation of the head does not seem enough. However, when you rotate your trunk you are inclined to say that a given object that was e.g. on the left before is now on the right. Thus, from a phenomenological perspective, it seems intuitively plausible to suppose that in normal cases the direction of a perceived object is indexed relative to the frontal axis of the trunk. The question then is whether this can possibly be established solely by proprioceptive signals from the neck upwards.

However, even if this can be made intelligible, opponents of the relationist view can cite another example that is taken to be a more severe case of deafferentation, namely "Christina", the 27-year-old "disembodied lady" described by Oliver Sacks (1986), Ch. 3.[14] Sacks describes this patient as suffering from a *complete loss* of proprioception due to the damage of nerve fibres that transmit proprioceptive signals following a polyneuritis. Nevertheless,

14 Note that the case is somewhat problematic, since Sacks's description and interpretation that there is a complete loss of proprioception has not been confirmed by others.

like IW, Christina managed to learn compensatory strategies and became capable of living on her own and managing her daily life. Nothing in Sacks' description speaks to the presumption that she has no egocentric spatial representations anymore.

One might even turn this proposal around and say that it is not the case in healthy subjects that there is an explicit relational spatial content in egocentric representation where one *relatum* is the subject itself, but *only* in pathological cases like this. From this perspective one would hold that IW's and Christina's pathology makes it necessary for them to always explicitly represent the position of their e.g. arm in relation to the target object to be able to interact with it. However, this is exactly the way we identified above as being certainly not the sense in which the claim is to be understood that egocentric spatial contents are relational.

3rd response: change of subject matter

A third way in which proponents of the monadic view might argue against the idea that proprioception grounds the singular representation in the perceptual content is simply to claim that this way of securing the relational structure effectively changes the subject matter. Since, to make this idea intelligible, one has to leave the level of pure visual perceptual representation and enter the level of this supposedly higher order state where representations originating from different sensory modalities are integrated.

Overall, I think that especially the phenomenology of egocentric perceptual representation supports the monadic view. The basic description we are inclined to give of the way in which the location of objects is presented to us in perception is in terms like 'there', 'to the left' or 'ahead'. Of course, based on these perceptual contents, we are also inclined to think or judge that the object is 'left from me' etc.; but this is an additional step that builds on the initial monadic presentation given in spatial perception. The objections discussed above are not sufficient undermine this picture. Furthermore, I think that in light of these phenomenological observations the two other reasons for endorsing the monadic view already given above are strengthened as well. Reflection on the phenomenology of visual perception motivates the attempt to specify the contents of egocentric representation by citing monadic spatial terms. Perry's work shows that doing this is possible without losing essential features we regard as linked to egocentric representations such as their immediate links to action (I will discuss his account in detail in sections 1.2.3 and 1.3.1). Furthermore, to the plausible phenomenological considerations and the verification of the possibility one can then add ecological considerations stating that this analysis raises the plausibility that more primitive cognitive systems are capable of egocentric representations as well. Finally, as a nice feature to have, it is also justified to indicate the greater theoretical simplicity as a feature, which would, standing alone, have not much import.

1.2.2.4 Constitution vs. content

I think the most fundamental way in which one can resist the claim of the defender of the relational view is to contrast the content of perceptual states on the one hand and the constitutive conditions these states depend on, on the other hand. The relationist's motivation for insisting on some explicit self-referential element in the content of egocentric representation is supposedly grounded in the conviction that only in this way it is possible to account for the action-guiding role that egocentric representations have. The question then is whether this role can also be explained without supposing the egocentric content to be relational. Although I take it that I did not present a completely convincing argument in favour of the latter thesis yet, there are at least some compelling points in favour of that direction. If it were the case that there is a good explanation of the action-guiding role of egocentric representations without relying on relational representations, one has another good reason to claim that egocentric spatial representations are actually monadic in nature.

The relationist thinks that the action-guiding role of egocentric representations can only be accounted for if both *relata*, the location of the object and that of the subject, are explicitly part of the content. That is, in effect, the subject not only has to represent the target object but the object that serves as the point of reference in relation to which the object's position is represented, namely the subject's body as well. A defender of the monadic view, in contrast, would hold that it is not necessary that the subject represents the reference object, itself, in any way to ensure that the represented spatial relation is indexed to itself. The idea would be that since egocentric spatial representation in general specifies spatial relations with respect to the subject, it is not necessary that the object of reference be represented in every particular case. The general consideration behind this claim is the following. To represent something as standing in some relation to oneself, one does not have (to be able) to represent the condition under which this representation would be correct, namely, that the object in question stands in some relation to another object, that is one's body. Although this latter condition is necessary for the representation to be correct, this fact in turn does not have to be represented by the subject.

A consideration of Perry's in a postscript to his "Thought without representation" might help to bring this out (Perry 1993a; cf. Perry 1998b, p. 87). There, Perry claims that it is not uncommon that we represent relations that are in fact e.g. ternary by representing a mere twofold relation (n+1-ary relations can be represented by an n-place predicate). His example refers to time. Suppose you are in a stadium watching ice hockey for the first time. When a team hits a goal there is a light flashing at the backside of the goal to indicate that the puck entered the goal. You as a spectator of the match represent the flashing of the light and the entering of the puck into the goal as two events that happen simultaneously. You would be inclined to judge that both events happen simultaneously. It appears that the content of this judgement may be expressed by the proposition "that the flashing of the light and the entering of the puck into the goal happen simultaneously". On the surface, it seems that there is only a twofold relation represented in that case, namely a relation between the entering of the puck and the flashing of the light. However, as Perry makes clear, the judgement is only true since "those events were simultaneously relative to my certain frame of reference" (Perry 1993a, p. 221). That is, a condition on the truth of such a statement is that the points in time of the events are represented in the same frame of reference. However, this fact itself does not show up in the content of the judgement.

An analogous example for representation of space can easily be found. Take for example Campbell's "Oriental despot". He has arranged things in a way that "however and whenever he moves, there is always Turkish delight just within reach and to the right" (Campbell 1994, p. 18). For this despot, the Turkish delight is at any point in time at the same place, namely, just within reach to the right. He would be inclined to say, "There is always Turkish delight to the right offered to me", e.g. when proudly presenting his empire to a friendly despot. Although this utterance is right in a sense and also comprehensible to the friend, it relies on a hidden element, namely the egocentric frame of reference of the despot. Thus, we have here a case where a twofold relation is expressed by a one-place predicate ('to the right' for 'to the right relative to my frame of reference').

Now, the crucial task for a defender of monadic spatial representations is to explain how exactly it is possible that egocentric spatial representation always specifies spatial relations with respect to the subject given that this is not reflected at the surface of the content of the representations. Just to claim that it is possible to express ternary relations by twofold relations does not explain how this is possible in the case of egocentric space. A convincing account of how this case is possible would give us a reason to claim that the contents of egocentric representation are in fact not relational in character.

This in turn has implications for the general question about the role of self-representation in spatial cognition. The view of the relationist was appealing because there it was claimed that egocentric spatial representation includes some singular reference to the subject itself, which made it straightforward to regard egocentric representation as a subclass of self-representation. However, if the arguments in favour of the monadic structure of egocentric representations are convincing, it becomes dubious whether these then can count as self-representation still, since then there would be no explicit representation of the subject part of the game. The question that emerges, therefore, is whether self-representation is possible without explicit representation of the subject.

Within the next section, I will start to tackle these issues. In particular, the aim is to demonstrate the possibility that egocentric representation can be seen as self-representation although it does not include any explicit reference to the representing subject in its content. In the last section of this chapter, I will introduce the notion of self-representation in more detail. I will consider two approaches, one that takes self-representation simply as a representation that explicitly refers to the representing subject and the other that takes self-representation as dependent on self-consciousness. I will suggest that both of these approaches are misguided. This will suggest that the notion of self-representation does not imply any explicit reference to the representing system itself. This in turn opens the possibility to regard egocentric representation as self-representation despite its monadic spatial contents.

1.2.3 'Self-representation' introduced

In answering the question what is distinctive of self-representation, the first idea that naturally comes to mind is the following: self-representation as a kind comprises mental states that represent the representing subject itself. Call this proposal 'Reflexivity'. According to 'Reflexivity', every representation that is of the representing system itself counts as an instance of self-representation.

Despite of the simplicity and intuitive force of this proposal, one can easily see that it cannot be adequate. Well-known examples from Ernst Mach and John Perry indicate that it is possible to represent oneself without the representation counting as an instance of self-representation. Ernst Mach tells that when he once entered a bus and saw a man coming towards him he thought, "what a shabby pedagogue is entering there" (Mach 1922, p. 3, n. 1; cf. Perry 1998b, p. 93). Unbeknownst to him, he saw himself in a mirror at the end of the bus. Thus, in this situation he represented himself without noticing that he did so. According to 'Reflexivity', this representation would count as a self-representation. However, this is somewhat odd since this representation does not have any distinctive properties compared to a representation of any other person in that situation. In that situation, Mach represents himself in the same way as he would represent any other passenger of the bus. Thus, although his representation in fact refers to himself, this does not make any difference. What significance could it then have to speak of an instance of self-representation in this case? None at all. It seems that mere factual reference is not enough to capture what is distinctive of self-representation. Perry's tale of the messy shopper points towards the same direction.

Perry's shopper who follows a sugar trail on the floor through the aisles of a supermarket, searching for "the one who is making a mess", not realizing that he himself is the person in question, is quite famous (Perry 1993b). The shopper represents 'that someone is making a mess' and, since he is in fact that man, his representation refers to himself. Thus, according to 'Reflexivity', this representation would count as an instance of self-representation, too.

Why should we be reluctant to follow this proposal? I think for two reasons that both highlight ways in which mere reflexive representations do not have the right place in the broader cognitive life of the representing subject to count as self-representation.

The first reason is that the representation in question is isolated from other instances of self-representation in the described scenario. Suppose the shopper knows that he has loaded his shopping cart with some bananas and a bottle of wine. Thus, we can say that the shopper now represents of himself 'that he is making a mess' and he represents of himself 'that he has loaded some bananas and wine in his shopping cart'. One might be tempted therefore to say that both representations are instances of self-representation. However, there must be some important difference between these two representations from the perspective of the shopper: If someone would ask him which things the messy shopper he is searching for would have in his cart, he would presumably answer that he had no idea and that he at most could infer from the sugar trail that he must have had some sugar packages in his cart. But how is this possible, given that we just said the shopper represents of himself that he has wine and bananas in his cart and that he represents of himself that he is making a mess? If these representations really would equally be instances of self-representation we would expect that the shopper knows what the messy shopper has in his cart, namely bananas and wine. However, he does not. He represents of himself that he is making a mess without knowing that this representation refers to him. He represents the fact in the wrong way.[15] That is why he would not know the answer to the question of what the messy shopper has in his cart. It seems very natural to suppose that contents of different instances of self-representation can be integrated in the sense that if someone represents of himself that he has some bananas and wine in his shopping cart and that he is making a mess, it should be clear that the right answer to the question would be affirmative and that he knows it to be so. In effect, we would expect all instances of self-representation to be known to apply to one and the same object in the world, namely to oneself.

The second reason for denying that the shopper's representation of the one who is making a mess should count as an instance of self-representation alludes to the fact that the representation is not apt to guide actions in the right way. This in turn might be brought out in two different ways. One way is more general, the other rather special. Together, these considerations will suggest that neither a representation *de dicto* of the shopper nor representations *de re* of the shopper are sufficient to ensure the special relevance for action that self-representations have.

The general way is this. Suppose the shopper, on his search, turns around a corner. Following the sugar trail with his eyes, he sees it ending at a location where two other shoppers with their carts stand. To fulfil his plan to tell the messy shopper that he is making a mess he would have to identify or single out the messy shopper first. Thus he needs not only to represent 'A is making a mess' but also 'this [one] = A'. That is, he would need some means of figuring out which particular person in the world his representation with the content 'the one who is making a mess' would apply to. This can be done in this case by means of perception: he just needs to look more carefully where the sugar trail ends or in which of the carts the damaged package of sugar is

15 Perry says that the subject has in this case merely "knowledge of the person one happens to be"; cf. Perry (1998b), p. 94; Perry (2011).

located. However, one thing seems clear: in order to tell the one who is making a mess that he is making a mess, the shopper needs to single out the person from all the objects and persons he can perceive. Thus, to explain his going towards one of the two shoppers and telling one of them that he is making a mess, we have to assume that he had in addition to his representation 'that A is making a mess' some representation of that man whose content might be 'this one = A'.

Thus, the first way of explicating that the original representation, 'the one who is making a mess', cannot have the right significance for action is by stating that the representation does not single out the 'target object' in the right way. One might formulate this by applying the *de re / de dicto* distinction to representations. It is not enough that the shopper represents that there is someone who is making a mess (*de dicto*) but in order to tell him that he is doing so he needs to represent of some particular person he can identify among all the other persons around as the person who is making a mess (*de re*). The most simple and straightforward way to do this is by some demonstrative representation.[16] It is worth noting that this point is not distinctive of self-representation. It applies to other representations as well. We can represent facts and properties about objects in the world without thereby knowing, or being able, to identify the referent of the representation even when perceiving it. What is usually needed for the latter is the representation of some distinctive feature and the recognition of that feature as had by some object that is perceptually represented. For our case this means that the shopper not only represents 'that someone is making a mess' but also that he notices that this representation applies to an entity he currently perceives. Only then does he know which entity in the world his former representation applies to. And only then is he able to act on that entity.

The second, special way in which it is brought out that the representation in question does not have the right significance for action concerns instances

16 The possibility of demonstrative representations is premised on the ability of the representing subject to locate the referred object in space; cf. Evans (1982). For the notion of *de re* representation see Hofmann (2001), who discusses and supplements the account given in Dretske (1995).

of self-representation in particular. For a representation to be an instance of self-representation, we would intuitively assume that it is significant for actions that concern the representing person herself. Take the example of a representation that 'S's pants are on fire'. We would only regard this representation as an instance of self-representation if the subject would react in an appropriate way. We would expect that he tries to extinguish the fire on *his* pants or tries to get off *his* pants.

We shortly saw that the shopper needs to have some *de re* representation of the messy shopper in order for his actions toward that shopper to be intelligible. The question now is whether given the assumption that the shopper would have such a representation of the messy shopper (of himself, in fact) that guides his actions we can regard this representation as an instance of self-representation. That is, the question is whether a *de re* representation that is in fact of the subject itself is enough to count as an instance of self-representation. My answer will be negative.

Suppose a slightly different scenario. The shopper again turns around a corner and suddenly and unequivocally perceives the messy shopper. This time it is clear that the perceived shopper is the one who is making a mess, since the shopper can see the sugar pouring out of one of the packages in that man's cart. Thus, the content of his representation of the man might be expressed as "*This* one is making a mess", instantiating a representation *de re* of the messy shopper. According to the considerations above, we do have a clear case of a representation that is directly relevant for the actions of the shopper. He would know how and where he would have to move if he wanted to tell the messy shopper that he is making a mess. However, unbeknownst to the shopper, the person he sees is actually he himself. There are some mirrors arranged in a way so that he perceives the messy shopper from the back, therefore not realizing that he sees himself. In this situation we clearly have a case where we have a *de re* representation of the shopper himself.[17] In addition, this representation is potentially action guiding with respect to that shopper. Since the shopper visually perceives himself and perceptual representations

17 Cf. Mach's example of representing himself as a "shabby pedagogue" above.

in general are those representations that paradigmatically are used to guide one's actions, the shopper would be able to act with regard to the person he perceives. Therefore, this case is a representation that is of the subject himself and that potentially gives rise to self-relating kinds of action. Do we therefore have an instance of self-representation?

The crucial question here is whether the action that the subject is inclined to perform based on this representation can be described as an action "that concerns the representing system itself", as we stated above. In one sense, this is clearly the case since the activities the shopper would perform to tell the messy shopper that he is making a mess are directed in fact at himself. However, on the other side, it seems clear that the shopper is completely unaware of this fact. The description under which he would perform his action is something like "to tell *that man* that he is making a mess". He would be very surprised when he would come to know that he himself is the one he sees in the mirror and the one who is the messy shopper.

In effect, what is shown thereby is that the proposal that instances of self-representation comprise *de re* representations of the subject that are significant for self-concerned action is still not precise enough – since these conditions are fulfilled in the just envisaged case without implying that the representation that drives the behaviour is a self-representation. The reason lies in the way the shopper represents himself. Although now it is a *de re* representation of himself, and there is an action performed with regard to himself, it is still possible, from the perspective of the representing subject, that the content of the perceptual representation and, accordingly, the description under which he would undertake his action, is not grasped as being about himself. This is just what happened in the mirror case. The subject would represent of himself that he is making a mess and he would be able to guide self-concerned actions based on this representation. However, it is likely based on the description of the situation that something crucial is still missing, namely some appreciation on the side of the subject that he represents *himself* as making a mess. Without this appreciation, the case is indistinguishable from any other case where some representation *de re* of an object guides actions toward that object.

One way of dealing with this problem would be to argue that genuine instances of self-representation are only in place when the expression of their content includes some form of the first-person concept 'I'. As soon as the shopper can be said to represent something like 'I am making a mess' all further questions as to how to stop it can be straightforwardly resolved. The shopper would immediately know what to do and he would know of whom in particular he knows that he is making a mess, namely himself. All the open questions from above would be settled.

1.2.4 Summary of Ch. 1.2 and outlook on Ch. 1.3

We started with a notion of self-representation that states that it is sufficient for a representation to be counted as an instance of self-representation when it is of the representing system itself ('Reflexivity'). This was unconvincing for the two reasons that there are cases of this type that are not linked to other instances of self-representation properly, and that there are cases that lack the practical significance we would expect from representations that count as self-representation. In further advancing the idea, we came to the thought that it must be at least some representation *de re* of the representing subject, since only these seem to be able to account for the relevant significance for action. But even this seems insufficient: there are cases of such representations of the representing system that are significant for self-concerned action but it still seems that they cannot count as self-representation because the subject is not in a position to grasp that it is a representation of himself. This seems to be only secured when we add the first-person concept to the content, which in turn speaks to the conclusion that self-representation comes only with the possibility of entertaining the concept of 'I', or, in other words, self-representation is nothing different from self-consciousness.

I do not want to embrace this line of thought, however and instead want to argue that there is an interesting and theoretically fruitful notion of self-representation that does not depend on self-consciousness. The kind of representation I envisage is situated between mere reflexive representations and representations that essentially include the first-person concept. The most important reason for supposing such an intermediate kind of representation is that I think there are scenarios that are best explained by alluding to a kind of representation that is tied in a special way to the representing system despite it seeming unnecessary to assume that mastering the first-person concept is necessary in these cases. The prominent case I will draw on is egocentric representation. I take it that the crucial difference between self-representation and self-conscious representation is that the latter includes some theoretical grasp, some understanding of the fact that what one represents is true of oneself, whereas in the case of self-representation the self-relatedness does merely show up in the peculiar way that instances of this kind are dealt with by the representing system. That is, instances of self-representation are treated by the representing system in the right way, namely as pertaining to it itself, without this being understood by the system in a theoretical or abstract manner.[18]

One crucial shift in this new approach to self-representation will be the following. Until now I tried to explain the peculiar roles that self-representation are supposed to play in the cognitive life of the subject by supposing that the representation is of the representing system itself. However, we saw that this does not guarantee the representation to fulfil these distinctive roles. Therefore, the new approach will be different. It will not try to ground the explanation of self-representation in the content of these representations, in particular not in the fact that the representing system itself features in the content of these representations in any way. Rather, the approach will take the distinctive functional role that these representations play as a starting point.

In the remainder of this dissertation, I will argue that although it is plausible to cast the content of egocentric representations in monadic terms, egocentric spatial representation nevertheless is as a subclass of self-representation. The reason will be that although the representations cannot be properly said to be 'about' the subject, they can be regarded as 'concerning' the subject, in that they are instances of 'agent-relative' representations (Perry 1986; 1998b;

18 The question about the relationship between self-representation and self-consciousness will not be pursued any further in my dissertation.

2011). This can be established by focussing on the conditions on which egocentric representations are constitutively dependent.

In the next chapter, I will present a first argument for the self-representational nature of egocentric representation that takes monadic spatial contents for granted. The argument is based on claims about the design of the cognitive machinery that underlies egocentric representation.

1.3 Egocentric spatial representation as self-representation II: Argument from design

In this chapter, I want to turn to a first argument that takes egocentric representation as self-representation while holding on to the monadic character of spatial contents. The proposal I want to discuss here may be regarded as an 'argument from design' of the cognitive machinery.[19] It relies on claims about the way that relevant information is processed by the cognitive machinery. I will first map out the argument in three steps and discuss each in turn. Basically, the argument purports to show that, on the one hand, there is not any explicit element to be assumed as part of egocentric contents that refers to the representing subject (monadic view about content) and, on the other hand, that egocentric representations are to be regarded as instances of self-representation nevertheless because of the functional properties these representations have. The discussion of the argument, especially that of the second step, will involve a detailed view on Perry's "Z-landers", a community that likes and succeeds in talking about the weather without ever mentioning the place they live in. One of the crucial questions here will be whether this case is sufficiently analogue to the case of egocentric representation.

19 The term is inspired by Anderson and Perlis (2005); cf. Beckermann (2003); Perry (1998b).

I will argue that the analogy holds and thus that egocentric representation can be regarded as 'concerning' the subject without representing it explicitly. I will start with Perry's general idea that higher-order relations (involving n+1 *relata*) can be represented by means of representing lower-order relations (involving merely n *relata*). After that, I will argue that this analysis can be applied to the case of egocentric representation. This will include a discussion of Perry's distinction between 'hidden *relata*' and 'global factors'. I will argue that the latter notion is the one to be applied to egocentric representation. Thus, egocentric representation concerns the subject since the subject functions as a global factor for this kind of representation. This shall explain how egocentric spatial representations can have monadic contents and nevertheless concern the representing subject.

In general, I think that the argument from design is on the right track in bringing out the self-representational character of egocentric space since it does not rely on the contents but on the constitutive and functional conditions of egocentric representations. However, in one crucial sense, the argument fails: it gives no appropriate account of the constitutive conditions of egocentric representation. Providing such an account will be my project in the chapters beyond Ch. 1.3. Thereby I present a criticism of the argument that might be understood as an undermining of one of the crucial assumptions the argument from design relies on. However, this does not mean that the gist of the argument from design is not worthwhile. In Ch. 4.1, I will argue in a similar way for the self-representational nature of egocentric representation but my argument will be based on a much more profound basis, namely on an account of the nature of egocentric representation that takes the actions of the subject as being among the constitutive conditions of this kind of representation. However, let me first start with the discussion of the argument from design.

1.3.1 The argument from design

The argument comprises essentially the following claims. (1) The cognitive machinery that grounds egocentric spatial representation is designed in a way that only ever delivers information that is specified with respect to the representing system. (2) Because of (1), there is no need for an explicit singular element in the content of this kind of representation. (3) Despite (2), it is possible to regard this kind of representation as a self-representation because of the peculiar functional role that representations of this kind play. – I will discuss these three claims in turn.

1.3.1.1 1st Step: Peculiar design

Let us start the discussion with claim (1), stating that the cognitive machinery that grounds egocentric spatial representation is designed in a way that only ever delivers information that is specified with respect to the representing system.

What is meant with this claim might be best brought out by considering some analogy. Take for example proprioception. Michael Anderson and Donald Perlis (2005) claim that "by design" (p. 307) the information that is present in proprioceptive states is about one and only one object, namely the representing system itself. This seems straightforward since the information channels that gather the relevant information all end beyond the skin as it were. It is their natural function, as one might put it, to deliver information about the state of the body, the position and movements of the limbs. It seems that it is impossible that they could ever collect information about a body other than that of the subject. Hence, it appears to be safe for the representing system to regard any information that is delivered via these distinctive ways as pertaining to its own body. A vivid example for such a hard-wired connection between the deliverances of this distinctive proprioceptive system and representations pertaining to the subject itself might be seen in cases of phantom limbs. Although known by the subject to be missing, there is the strong impression that an experienced limb is still there. This might be explained by the fact that information channels whose natural function is to provide information about the limb are partially intact still. Therefore, any information they deliver is the basis of representations of a limb that is not actually part of the body anymore.[20] Whatever information comes through these channels will give rise to representations of own body parts.[21] This indicates that what is decisive is not the object from which the information originates but only the kind of information channels that grounds proprioceptive representations.

Can this reasoning be applied to the case of egocentric representation? I think the analogy between the two cases rests on the function of the information channels that ground each kind of representation. The function of proprioceptive information channels is to deliver information about one's own body. The analogous function of egocentric information channels would then be to deliver spatial information about external objects where this information is specified with respect to the subject. That is, information you receive via these distinctive channels leads to representations with egocentric spatial content. Thus, e.g., everything you hear, you hear as coming from some direction relative to you. Everything you see, you see as being located in a certain distance and direction from you (Beckermann 2003, pp. 179–80). In contrast, you do not smell things as coming from a distinctive direction and distance relative to you, at least by far not as precise as in the case of vision and audition, although it is certainly intelligible that this was the case, and it is apparently in other mammals.[22]

The first step of the argument from design thus consists in making clear that the contents or essential properties of contents of the representational states in question are determined by characteristics of the distinctive ways the

20 This example is due to Anderson and Perlis (2005), pp. 311–2. Note that Anderson and Perlis use this argument from design to illuminate the idea of self-representation as a singular representation of the representing system. That is, they would not underwrite claim (2) above. Beckermann (2003), who presents the same kind of argument, in contrast, clearly endorses (2).

21 One might even think of strange cases where someone's proprioceptive channels are wired up with someone else's body or with an artificial body. On this account, the subject would regard the information gathered in this way as pertaining to its own body.

22 Experimental results on spatial information in human smell are not that unambiguous, though. It seems that there is some rudimentary ability to decide whether some smell comes from the right or left; cf. Porter et al. (2005); Sela and Sobel (2010), p. 17; Békésy (1964).

information, which is the basis of these representations, is processed. For the case of proprioception this means that all information gathered by the ways that normally transmit information from the own body is treated as pertaining to the own body, and is therefore the basis of proprioceptive representations, even in cases where there is no actual referent, e.g. in phantom limb cases. For the case of egocentric representation this implies that information that is transmitted in ways that normally give rise to egocentric representations, such as audition or vision, will lead to representations whose spatial contents are specified with respect to the subject, even in cases where there is no adequate cause of the information (hallucination, or illusion cases). On this picture, the way information is transmitted (partly) determines the characteristics of the resulting representations, in our case the property of egocentric representations that their spatial contents are indexed relative to the subject.[23]

In this sense, claim (1) says that it is the characteristics of the information processing machinery that ensure that egocentric representations only ever specify spatial magnitudes with respect to the representing subject. In the next step, I will discuss claim (2) that holds that in virtue of (1) an explicit element in the content of egocentric representation that refers to the subject is not necessary.

1.3.1.2 2nd Step: "Concern"

Let me turn to claim (2). It states that since the information gathered via the ways specified in (1) by their natural function always pertains to the subject, there is no need to assume that the subject needs to explicitly represent itself as being the one to which the information pertains (Beckermann 2003; Perry 1998b).

This step is not so straightforward. *Prima facie*, there is a good alternative to endorsing (2), expressed in Anderson and Perlis (2005). They suggest that

23 Below, I will present a critique of this way of spelling out the constitutive conditions of egocentric representation. As a response, I will develop a much more substantial picture of the nature of egocentric representation that relies on the general idea that action is a constitutive condition on spatial perception. One could take the argument that I will present as backing up the claim that is made here, namely that egocentric representations *qua* kind are indexed to the representing subject.

since the cognitive machinery ensures that every representation that is based on these ways pertains to the representing subject, the subject can safely predicate the content of these representations to himself. In effect, they see in this necessary self-reference a secure basis to hold that there is an explicit element in the content of the representation referring to the representing subject itself. This shows that Perry's and Beckermann's claim that there is no need for an explicit element in the contents of egocentric representation since it only ever specifies spatial relations with regard to the subject needs an argument.[24]

Perry argues that egocentric representations do not have to explicitly refer to the representing subject in order to account for the causal and computational roles that the representations are supposed to have. Perry starts with the premise that the ways of gathering information in egocentric representation are *per se* of the representing organism and that it is not even possible that they are about anything else. This is basically what I discussed as claim (1) above (Perry does not explain this fact, and neither does Beckermann). He argues that it is therefore not necessary to assume some singular element in the contents of egocentric representations that refers to the subject. The general line of thought seems to be that an explicit element in content is only needed where there is some range of possible referents that the content might apply to. Since this is not the case for representations such as egocentric representations (since their spatial contents are always indexed with respect to the subject), these are better described as lacking any singular expression that refers to the subject explicitly.

I take it that step (2) of the argument from design is the crucial point. It has to be made clear why an account of the contents of egocentric representation need not draw on any singular element referring to the subjects of these representations, i.e. to show that a monadic view of the contents of egocentric spatial representation is plausible. In what follows I will therefore discuss this step in detail. This discussion will largely focus on Perry's imagined people of Z-land and their way of talking about the weather. Perry uses this example to

24 Note that Beckermann (2003) does not make explicit this argument and simply quotes from Perry.

make clear how some content can be about a certain subject matter without even the possibility of this subject matter showing up explicitly in the content. I will first develop the case of the Z-landers, which already includes an important distinction between two ways in which some content can be about something in an implicit way. The contrast is between what Perry calls 'unarticulated constituent' and 'global factor'. I will make clear that it is the latter notion that we need, exemplified by the Z-landish way of talking about the weather. Later on, I will discuss the crucial step of how the analogy between this case and the case of egocentric representation is to be understood.

Unarticulated constituents and global factors: Weather discourses in Palo Alto and Z-land

I already discussed the reasons that speak in favour of Perry's claim that the contents of egocentric spatial representation are monadic, i.e. that they do not contain any self-referring element (see section 1.2.2 above). In this section, I want to explicate in detail the sense in which the singular element referring to the subject need not be part of the content of egocentric representations. This will make clear in what way claim (2) has to be understood ultimately. Perry distinguishes between two different ways in which the idea could be understood that there is something that does not actually show up in the content itself but that has to be taken into account when assessing the meaning of the content of some representation: either there is an "unarticulated constituent", or a "global factor" (Perry 1986).[25]

The first of these notions can be explicated by considering one of Perry's examples. While he is lying in his bed, Perry hears his son saying, "It is raining". Perry treats this remark naturally as an expression of the fact that it is raining in Palo Alto, where he and his son currently are. He therefore abandons his plans to go playing tennis and stays in bed instead. Thus, the assertion is understood and treated as to be true if, and only if, it is currently raining in Palo Alto. However, there is no element in the content, no "component" of the content, which refers to Palo Alto. Nevertheless, Palo Alto clearly has

25 Perry (1998a) further divides unarticulated constituents into context-sensitive and context-insensitive kinds. However, this distinction is not important for my discussion.

to be taken into account when assessing the truth of the statement. Thus, Perry concludes, Palo Alto is an unarticulated constituent of the assertion. The statement is in effect *about* Palo Alto without actually referring to it explicitly.

We will see that this notion, although it includes the feature of being about something without explicitly referring to it is not enough for the argument from design, since for this we need a more fundamental way in which egocentric representations are about the subject without explicitly referring to it than is instantiated in cases that have some unarticulated constituent.

We find this more fundamental way realized in Perry's notion of a 'global factor'. *Prima facie*, this notion seems to be indistinguishable from the notion of the unarticulated constituent. However, on closer examination, it is not as we will see shortly. Perry introduces the tale of Z-landers, "a small isolated group, living in a place we call Z-land" (p. 144), to make clear the difference and spell out the peculiarities of the 'global factor'. What is important about Z-landers is that they never leave their small territory, they have no contact to people outside of Z-land and they even have no name for their own country. When one of the Z-landers utters the statement "It is raining", it is clear for everyone hearing the statement that it is currently raining in Z-land. When someone says "It will be raining tomorrow", everyone would be inclined to adjust their plans for the next day. For everyone, it is clear that statements about the weather pertain to the weather in Z-land. And this is so without any explicit mention of Z-land at all. On the face of it, "Z-land" seems to be just an unarticulated constituent as "Palo Alto" was in the example above: the truth of the statement depends on facts about Z-land although there is no component in the assertion that refers to Z-land.

However, Perry claims that there is a crucial difference between the two cases. The difference between the two notions lies in the role of the context. Let us consider the role of the context in the case of the unarticulated constituent.

When his son tells Perry that it is raining, it is the context of the assertion that determines the unarticulated constituent; in this case 'Palo Alto'. In general, our assertions of the kind "It is raining" can, in principle, be about many different places. Which place every instance is about is determined by the context of the use of the phrase, in particular, by the intention of the utterer to refer to some specific place. In principle, there is always an open question in relation to which place the assertion has to be assessed as true or false, although in most times of our daily lives we do not have to reflect about the place in question since it is straightforwardly clear from the context. Nevertheless, there are also situations where misunderstandings can happen.

Having just finished a phone call with someone in London, your friend tells you that it is raining. You might have forgotten that she just was on the phone and understand the utterance as pertaining to where you currently are, whereas your friend meant to indicate that it was raining in London. Another example would be that you are on a long car ride, you slept for some hours and when waking up the driver tells you that it is raining. You would think that it is raining in Germany, whereas unbeknownst to you, you are already in France. These are cases where the hearer of the utterance assesses the statements based on the wrong context.

The situation is different in the case of global factors. The just described scenarios would not be possible in Z-land. Z-lander's weather discourse, as one might want to say, is always and ever only about exactly one place, Z-land. By hypothesis, Z-landers do not know of weather statuses in other places, they even have no idea of places other than Z-land at all. Thus, one might illuminate their situation regarding weather conversations in saying that the possibility space that consists of all possible places about which a given weather assertion could be made includes just one place, namely Z-land. That is, every statement about the weather *concerns* Z-land, without there being any component of it that refers to Z-land, or there being any ambiguity regarding which place is concerned that has to be solved by looking at the context. Thus, Z-landers communicate about properties of a certain place without ever having to mention the place about which they are communicating. 'Z-land' functions as a global factor of their weather discourse in the sense that it is a point of reference for any utterance about the weather. In their language, weather is only indexed to time, not to locations in space (Perry 1986, p. 146), because there is actually just one place for Z-landers about which utterances about the weather make sense. What additional explanatory value would it have to suppose that the Z-lander's beliefs about the weather would each have an 'unarticulated constituent'?

In contrast, as we saw before, in human weather discourse there are as many places in the possibility of space as there are places in the world. Since it is, in principle, always possible that assertions about the weather are about one of many different places, there is the need to be able to somehow identify the place in question, which is in most times simply done by considering the context of the assertion, that is, by way of an unarticulated constituent. If the hearer of an assertion would be uncertain as to which place the assertion is meant to be about, he could ask the utterer for an explicit statement about the location he was referring to. In Z-landish weather discourse, there cannot be such a situation, neither would the question make sense nor would there be any answer to be given.

The basic rationale behind Perry's line of thought then is that there is only the need of some singular representation of some entity, articulated or unarticulated, in case there are at least two objects of a certain kind that may be predicated with the same property.[26] For Z-landers there exists just one place and the discourse about the weather in Z-land accordingly only ever concerns the weather conditions in that very place. This is a fact that is grounded in the 'lifestyle' of the Z-landers, in particular in their ignorance of any place other than Z-land. Therefore, there is simply no need for some singular representation or idea of that place in the discourses about the weather. It is here where Perry's already mentioned basic principle about relations and their representations has its theoretical place (see above, p. 28):

26 "But I do not think that having a cognitive relation to a family of properties requires one to understand that there are a range of things that can have or not have the properties" (Perry 1993a, p. 222).

> The general point is this. Sometimes all of the facts we deal with involving a certain *n*-ary relation involve the same object occupying one of the argument roles. In that case, we don't need to worry about that argument role; we don't need to keep track of its occupant, because it never changes. We can, so to speak, pack it into the relation. (Perry 1998b, p. 87)

Statements about the weather normally are statements that describe certain conditions relative to certain times and places, that is, in a ternary relation. Since Z-landish weather discourse cannot be about anything other than Z-land, there is no need for an explicit reference to the place. The ternary relation is represented as a binary one.

To mark this distinctive way of reference, Perry says that the Z-lander's weather statements are not *'about'* Z-land, but that they *'concern'* Z-land (cf. Perry 1986; Perry 1998b; Perry 2011; Beckermann 2003). The notion of *concern* here means that Z-land has to be taken into account when assessing the truth of some statement about the weather although not being possibly explicitly referred to.

Now that we have the appropriate notion of 'hidden' reference in place, i.e. the 'global factor', we can go on and ask how egocentric representation is supposed to be understood as involving this feature.

The case of egocentric representation

The analogy to the case of egocentric spatial representation is as follows. As was explicated by (1), all spatial information that is represented egocentrically specifies spatial relations relative to the subject. There is no possibility that spatial relations are ever specified with respect to someone other than the representing subject. This is secured by the way the contents of egocentric representation are grounded in the information processing machinery. It follows that the analogy between the Z-landers and egocentric representation is supposed to hold in the following way: Weather statements in Z-land (as a kind of assertion) correspond to egocentric representation (as a kind of representation). Particular assertions about the weather in Z-land correspond to particular instances of egocentric representation. Z-landers' ignorance of other places than Z-land corresponds to the natural function of egocentric representations to represent spatial relations only with respect to the subject. The global factor 'Z-land' that is part of the truth conditions for every statement about the weather corresponds to the global factor 'subject' in the case of egocentric representation. In both cases, the global factor does not show up in the content of the statement, or representation but it is present in determining the truth conditions of the statement, or egocentric representation, respectively. Let me elaborate this point a bit.

To make the point more clear: the content of egocentric representations usually includes some part that specifies spatial information and some part that refers to something represented as being at that position, some object in the broadest sense (even shadows or sounds count as objects in this sense), e.g. 'x is to the left.'[27] What is certainly true of the content is that its correctness depends on the actual spatial relation between the represented object and the subject at the time of representation. This is the same situation as it is with the Z-landers. The truth of their assertions about the weather is also dependent on how the weather actually is in Z-land at the time of their assertion. In both cases, one essential element that is part of the correctness condition is not mentioned in the content of the representation or assertion, respectively.

One might wonder why we need the notion of global factor instead of the simpler notion of unarticulated constituent for the case of egocentric representation. The reason is this: If we would conceive of the missing singular element ('subject') as an unarticulated constituent like 'Palo Alto' was in the first example I discussed above, one might argue that it is possible that only the public assertion does not contain the singular element, whereas the underlying mental state of which this assertion is an expression does.

Take as an example the one we already mentioned where your friend, who just finished a phone call with someone in London, utters that it is raining. She might very well think that it is raining in London, but she only asserts the shortened statement, expecting that you would know from the context about

27 'x' is just a singular element that is taken to refer to a particular thing. It might be replaced by "this" or by "John" or by "the man I met last night".

which place she is talking. Thus, she in fact intends to speak about London, a place where she currently is not at, but she also would be able to tell you that she was speaking about London. If 'subject' would have been just an unarticulated constituent like 'London', it would implicate that it was not general true of egocentric representation that there is a global factor but only in certain cases. This would undermine the claim that egocentric representation *qua* kind concerns the subject without explicitly referring to it.

However, the peculiarity of the global factor is exactly that it functions as a point of reference that is relevant for the appropriateness of representations in any token case of some more general kind without this being grasped by the subject. The crucial insight from the example of the Z-landers is that contents can be about some object in the sense that their truth depends on it without this object being part of the content. The Z-landish weather discourse is an extreme case: the Z-landers do neither intend to speak about Z-land, nor are they in a position to ever make explicit that they are referring to Z-land when talking about the weather. They do not even have a name or notion of their country. Arguably, they have no grasp of the concept of place at all. Ultimately then, the status of 'Z-land' as a global factor in Z-landish weather discourse is based on the fact that the range of objects to which weather-related predicates are applied in their language and thinking is exactly: one, namely Z-land. No alternative objects to which a given property can be ascribed means no need for explicit representation. Thus, the global factor plays the role of a point of reference for some relation that is fixed for a given domain of representing. In the domain of thoughts and statements about the weather in Z-land, it is Z-land itself. In the domain of egocentric spatial representation it is the subject who has the representation.

Objection

I want to consider a possible objection to that proposal. Turning to egocentric representation raises the question whether it is actually appropriate to apply the notion of global factor to the case of egocentric representation. Are we *really* in the same situation with respect to egocentric representations as the Z-landers are with respect to their weather discourse? In particular, whereas their conversations about weather certainly only ever concern Z-land, without this being or even possibly being transparent to them, we of course can conceive of egocentric representations that are not our own. It is no problem for me to represent that something is left of you while being right of me. Hence, it appears we are perfectly able to make explicit the 'hidden *relatum*', to intentionally represent that something is on the left *of me* or to represent that something is to the left *of you*. We are not as limited in thought and speech as the Z-landers are. The set of possible objects to which we are able to apply egocentric spatial properties includes not only one, but rather infinitely many.

Why then is it legitimate at all to suppose that 'subject' is actually a global factor in human egocentric representation as 'Z-land' is in Z-lander's weather discourse? It seems the cases are not analogous. When there is the capability of representing the same property as had by two distinct objects, there is a need for the capability to represent the bearer of the property in a particular case explicitly. Z-landers lack the former capability and hence it is not necessary to assume that they have the latter. However, it seems that we as subjects capable of egocentric representation have the former capability – and thereby the latter, too. Thus, it seems that the analogy between the two cases does not hold.

The crucial point to which this sort of challenge leads is to suppose that 'subject' in the case of egocentric representation is at best merely an unarticulated constituent (since the hidden *relata*, the 'here' or 'me' that is part of the represented spatial relations can easily be made explicit) and that therefore the cases of the Z-landish weather discourse and egocentric representation cannot count as analogous. This would mean in effect that Perry's tale of the Z-landers cannot be used to argue for the claim that 'subject' really is a global factor in cases of egocentric representation.

However, I think the objection is unsound for the following reason. The objection relies on a wrong assumption. The assumption is that the spatial information in human egocentric representations can be indexed to other persons than the subject; but this is not the case. It is true that we can represent objects as standing in the same spatial relations to others as we can represent them standing relative to ourselves. We can represent that "the table is to the left of Z". However, and this is crucial, this sort of spatial representation is not egocentric, but allocentric.[28] That is, in claiming that we are able to represent that the same spatial relations hold between some object and someone other that hold between the object and ourselves, the subject matter about which we talk changes.

The easiest way to see this is by considering the fact that the representation that "the table is left to Z," is not immediately relevant for *my* bodily actions. At least I would have to rely on some translation based on the represented spatial relation and some egocentric representation of my spatial relation to 'Z' in order to be able to interact with the table. Thus, it clearly is not a case of egocentric representation at all, since as I argued above these are partly defined by their immediate relevance of action. There is always a gap between these other kinds of spatial representation and their use in action that has to be closed by some egocentric representation. Hence, contrary to what the objection states, representing "the table is left of Z" is not simply an egocentric representation with a hidden *relatum* stated explicitly. There is a different kind of spatial representation at work. Thus, the objection relies on an implicit change in the subject matter from egocentric representation to allocentric representation – and therefore, it is unwarranted.

The objection is grounded on a misunderstanding of the conditions under which it is possible for me to represent that "the table is left of Z". The capability for these representations seems to me to rely on the capability to 'put oneself in someone else's shoes' by means of imagination (perspective taking). In effect, this means that to be able to represent that "the table is left of Z" one has to be able to represent egocentrically (in imagination) that "the table is to the left". Thus, the egocentric representation itself is never indexed to someone other than me, contrary to what the objection states. What makes representations such as "the table is to the left of Z" possible is that humans are able to transcend their egocentric way of representation and apply the included predicates to other persons and objects. However, this constitutes

28 Literally, "allocentric" means "centred on someone else".

another form of spatial representation, in particular, allocentric representation. It is not something that is inherent in the egocentric way of spatial representation itself. It relies on transcending the egocentric way of representing towards another kind of spatial representation, namely the self-detached, allocentric way of representing space.

Ultimately then, when one regards only egocentric representation of space, humans are in the same position as the Z-landers are with respect to the weather. For egocentric spatial representation, there is not even the possibility that the information is indexed to someone other than oneself. Therefore, we need not make it explicit and a creature that would only be capable of egocentric representation could not do so. The fact that we are indeed able to represent others as standing in the same spatial relationships as we do (and to represent explicitly that we stand in certain spatial relations) relies upon a capability that lies beyond egocentric representation of space. The proponent of the objection above is ultimately mistaken in supposing that the crucial analogy is between humans and Z-landers, while it is in fact to be drawn just between humans in the egocentric mode of representation and Z-landers.

Summing up step (2)

In sum, I take it that Perry's analogy between the Z-landers and egocentric representations is justified. On this view, egocentric spatial representation 'concerns' the subject, although it is not 'about' it. The objection that egocentric representation involves at best an unarticulated constituent is not sound since it relies on an improper understanding of egocentric representation and of the proposed analogy.

Now we arrived at an analysis of the details of Perry's and Beckermann's account and the idea of the way in which egocentric representation is supposed to be self-referring without being explicitly about the subject. Starting in 2b) I developed the idea that egocentric representations can be analysed in monadic terms, i.e. without supposing that they explicitly refer to the subject in any way. I started with Perry's general idea that higher-order relations (involving n+1 *relata*) can be represented by means of representing lower-order relations (involving merely n *relata*). The task for this chapter was to make clear in what way this applies to the case of egocentric representation. I first distinguished two different manners in which one can speak of hidden *relata.* In the first manner, it is the context of some utterance or representation that determines the hidden *relatum* ('unarticulated constituent'), whereas in the second it is much more fundamental facts about a given area of representation that determine the hidden *relatum* ('global factor') for any instance of this area. I argued that if we follow Perry we have to regard the case of egocentric representation as one where a global factor, determined by the particular way that egocentric spatial information is processed, is in play. The global factor is the subject itself in these cases.

This is where we are now. Thereby, we have established steps (1) and (2) of the argument from design, stating that egocentric representations are only ever indexed to the subject and that it is therefore in order to regard these representations as lacking any explicit reference to the subject. The last step will be to make intelligible the idea that although the subject is not a component of the content of egocentric representation, neither articulated nor unarticulated, it is nevertheless intelligible and right to think of this kind of representation as a kind of self-representation. Thus, I now want to turn to step (3) of the argument from design.

1.3.1.3 3rd Step: Peculiar functional role

Claim (3) states that what makes egocentric spatial representations instances of self-representation (despite not being about the subject explicitly) is their peculiar functional role. The aim of this last section is to make clear how this works given the proposal of a 'selfless' egocentric representation elaborated in the former steps.

Egocentric representations have an irreducible place in the control and guidance of the subject's actions. The question arises whether it is possible to explain the peculiar functional role that egocentric spatial representations play based on the proposal embodied in claim (2)? We saw that it is precisely the fact that egocentric representations specify spatial relations relative to the subject that they can be used directly to guide the actions of the subject towards an object. However, it seems that on the present proposal the content does not actually specify the relation between subject and object. It merely includes the object and its properties, where one of its represented properties is a certain spatial property, e.g. "to the left". How should this be immediately action-guiding for a subject?

Maybe it will help to draw an analogy to the situation of the Z-landers again. Their representations about the weather also do not contain Z-land as a component. However, arguably, if a Z-lander represents that it is raining, it will grab an umbrella if she does not want to get wet outside. The representation can obviously serve the action-guiding role without explicitly referring to the place where it rains in these cases. The reason was, as we saw, that there is no possible open question as to where it rains, because every representation of the weather status is by itself a representation of the weather status in Z-land. Thus, the reason why it is possible that Z-lander's representations of the weather status can be action-guiding without being about (in Perry's sense) Z-land lies in the fact that every instance of weather representation of Z-landers will always concern the weather in Z-land.

Applying the analogy to the case of egocentric spatial representation would lead to the following picture. Egocentric spatial representations can be action-guiding for the subject although they are not about the subject, since their natural function is to specify spatial relations with respect to the subject. Therefore, any instance of egocentric spatial representation does concern the subject in that the conditions under which it is correct include facts about the subject. Representing something as being to the right is only correct if that something is to the right of the place where I actually am.

In effect, what I am claiming is that egocentric spatial representation can count as self-representation although instances of it are not about the subject, thus, they do not include some singular representation of the subject, either articulated or unarticulated, in the content. Perry's notion of concernment is enough to establish the kind self-representation. It follows from this view that what ultimately matters is not the object in the world the representation is about but merely the practical significance the representation has for the representing system.[29] We already saw that this practical significance cannot be secured even by the representation's being about the subject (cf. the messy shopper case above where he represents himself but not in the right way). On the other hand, any instance of egocentric spatial representation can immediately be put to use to direct one's actions. For example, it is enough to represent a ball as 'to the right' to be able to move based on this representation so as to orient toward it, approach it or grasp it etc. There is no intermediate step necessary, part of which would be some transition of the spatial information into something like 'to the right of *me*'. This is ruled out as necessary since the natural function of every system that delivers egocentric spatial representation is to specify the spatial relation with respect to the subject. Thus, one does not have to make out relative to which location one represents something as being 'to the right'. It does always already concern the subject.

What would make this approach impossible would be a successful argument to the end that one cannot explain the immediate action-guiding functions that egocentric spatial representations play without assuming that there is some element in the content of egocentric representations that refers to the representing system. The argument would in effect have to question the assumption that it is possible that a state that represents the binary relation between subject and object by representing a just monadic property is poised to guide the subject's actions toward the object. But it seems dubious to me how this should work. The reason is that the crucial information that is needed for goal-directed action is where the object is relative to the subject. And this information is already given by the monadic contents that are determined by the spatial relation between subject and object.

29 Beckermann would not endorse this claim. For him self-representation is linked to explicit representations of the representing system, as he told me in personal communication.

1.3.2 Critique of the argument from design

Although I think that the stated objection to the argument from design is not successful, I want to criticise this way of cashing out the argument from design. My basic point is that step one of the argument is either substantively underspecified or wrong.

As outlined above, the argument from design involves three steps. (1) The cognitive machinery that grounds egocentric spatial representation is designed in a way that only ever delivers information that is specified with respect to the representing system. (2) Because of (1), there is no need for an explicit singular element in the content of this kind of representation. (3) Despite (2), it is possible to regard this kind of representation as a self-representation because of the particular functional role that representations of this kind play.

The central claim on which the whole argument from design hinges is to be found in premise (1), namely that the cognitive machinery that grounds egocentric spatial representations is designed in a way that only ever delivers information that is specified with respect to the representing system. Stated like this, the claim is not false, but it is not correct either. The underlying problem consists on an unarticulated background assumption that is likely to be part of claim (1). The assumption would state that explicating the design of the information processing machinery is *sufficient* to explain the crucial property of egocentric representations, in particular, that spatial information in these cases is indexed relative to the subject. This claim can be understood in at least two different ways.

On a sufficiently wide understanding of "cognitive machinery", this is uncontroversial, but it explains hardly anything. If "cognitive machinery" refers to everything that is involved in the generation of egocentric representation, the claim is certainly true but not informative. It would be necessary to spell out in what way this claim is meant and which elements of the cognitive machinery are key. An explanation of egocentric representation cannot stop with pointing to the global design of the cognitive machinery.

However, if "cognitive machinery" is understood in a narrow sense as merely referring to the neuronal modules and processes that execute information processing of incoming information that eventually results in egocentric representation, the assumption has to be rejected as will be made clear in the subsequent discussion of the claim that action is a constitutive condition on egocentric representation. To hold that perception is constituted alone by the passive processing of visual input would mean to endorse what Hurley has dubbed the "Input-Output-Picture" (see Ch. 2.2.2). The discussion of the claim that action is a constitutive condition on perception will make clear that this model is inadequate. Perceptual egocentric space is constituted in part by the practical engagement of the subject with the world. This is the common conviction that is shared by the otherwise largely differing accounts I will discuss in Part 2 and the position I will finally defend in Ch. 4.1.

Therefore, if the argument from design rests on this narrow idea of accounting for the nature of egocentric perception simply in terms of the way visual input is processed, the argument is compromised by an inadequate understanding of visual perception. On the other hand, if it rests on the broad notion of cognitive machinery noted above it would suffer from being largely uninformative. It would not really give an account of why egocentric representation is self-representation since it would not really explain the constitution of egocentric representation but merely propose that this is what the self-representational nature grounds. Unless the argument is backed up with details about the way egocentric representation is constituted, it is unconvincing.

I take it that the overall strategy of the argument, i.e. focussing on the constitutive conditions of egocentric representation instead of the contents, is right. However, the constitutive conditions are different from the ones (possibly) envisaged in there, which will be shown in the remainder of my dissertation.

In the parts to follow, I will develop the claim that the self-representational character of egocentric representation is based on a sense of one's own practical possibilities that is a constitutive condition on spatial perception. I will start explicating this claim in the next Part. But first, let me summarize the arguments developed hitherto.

1.3.3 Interim conclusion and outlook

Let me recapitulate the discussion so far. In Ch. 1.2, I began by asking what is distinctive of egocentric spatial representation and how it can be understood as a kind of self-representation. The first proposal of an explanation for this was that egocentric representations are instances of self-representation since they include some *de re*, but not self-conscious, representation of the subject. Although this proposal would quite directly lead to the conclusion that egocentric representations are a kind of self-representation, it is unlikely, since there are compelling reasons to regard the contents of egocentric representations as monadic in nature (as given in section 1.2.2). I spelled out this approach by reference to John Perry's work and gave some reasons for it. In the remainder of Ch. 1.2, I turned towards the notion of self-representation. I went through some scenarios that seem to indicate that self-representation is to be conceived of as self-conscious representation. This in turn was marked as implausible, since there seem to be cases where there is good reason to assume involvement of some instance of self-representation but where it is not likely that the subjects are self-conscious.

In the present chapter, I considered the argument from design that does not only do justice to the monadic nature of egocentric representational content but also to the categorization of egocentric representations as self-representation. This argument essentially answered two questions. The first is, how is it that egocentric representations can have monadic contents (steps one and two)? The second, why are they still to be regarded as instances of self-representation (step three)? I discussed the details of the argument, specifically the proper understanding of how it is to be understood that egocentric representations concern the subject without representing it explicitly. I argued that of the two possibilities that are proposed by Perry, the notion of 'global factor' is the right one to be applied to egocentric representation. The peculiar functional role of these representations, consisting in their immediate action relevance, which in turn is based on the fact that each egocentric representation concerns the subject, ensures that egocentric representation is to be regarded as self-representation.

The following chapters can be seen as a critique of this argument, especially of its first step. The agenda will be this. I will present a historically informed and systematically ordered discussion of views that take action as a constitutive condition on spatial representation (Part 2).[30] In addition, I will discuss empirical findings that are relevant to the topic (Part 3). The discussion shall show that there are good reasons to suppose that spatial representation cannot simply be thought of as constituted by some peculiar passive way of processing information originating from the environment. A proper understanding has to take into account the active engagement of the subject with the world. There are several different accounts of this basic idea. A discussion of these approaches will set the stage for the development of my own view. The chapters on empirical evidence shall then help to further argue for the basic claim and to decide between the basic approaches. Finally, in Part 4, I will again turn toward the task of arguing for the self-representational nature of egocentric representation. The argument there will be based on a much better understanding of the constitutive conditions of egocentric representation. In addition, the intimate relation between a sense of one's own capabilities to move and egocentric spatial perception, which I will argue for, implies the self-representational nature of egocentric representation.

30 The primary *explanandum* will be visual spatial perception since it is the most discussed form of spatial representation.

2 Action

Why do humans perceive the world spatially? A glimpse at the history of philosophy reveals various different explanations. For example, while Descartes thought that the properties of the retinal image are central, Kant referred to space as a condition *a priori* of intuition. Later on, Husserl claimed that spatial experience depends on the actions of the subject. In this part of my dissertation, I want to focus on approaches of the latter kind. Many accounts that spell out the basic idea that spatial perception is dependent on the actions of the subject have been proposed. However, since action is a multi-faceted phenomenon, the accounts vary widely in how they spell out the basic idea. Some rely on actual movements of the subject, some focus on the capability to act. Others propose that spatial experience is grounded in the knowledge about the sensory consequences of bodily actions.

I want to give an overview about the different accounts that share the claim that action partly constitutes spatial perception. The purpose of the overview is to inspire and to set in context the view that I want to argue for in the end. There are three chapters in this part. In Ch. 2.1, I will not only present some of the historical development that the main claim has taken since the middle of the 19th century. I will also answer the question of what it means to posit a constitutive dependence between action and perception in the first place.

The remaining two chapters of this part are devoted to giving a systematic grouping of different kinds of accounts. The accounts differ with regard to the way action is taken to be constitutive, ranging from actual movements to mere knowledge about the sensory consequences of possible movements. The result will be a comprehensive overview about how the constitutive dependence of spatial perception on action can be cashed out. Some accounts will be criticised and rejected. However, the final decision about which of the discussed accounts is most appropriate will be reached only in Part 3 where I will present and discuss relevant empirical evidence to see which approach can explain the findings best.

In sum, this and the next part will bring us all the way to an answer to the question of whether and in what sense spatial perception is constitutively dependent on action. Given that we can assume this constitutive relationship, it will be possible, finally, to establish the claim that egocentric perception is a form of self-representation on a most secure basis.

2.1 Action as a constitutive condition on spatial representation: Introduction

This chapter serves two central aims. In the first section, I will present a rough overview about the history of the idea that spatial perception is dependent on action in one or another way. This will reveal substantial differences in how this claim can be defended. Before investigating the different kinds of approaches in a systematic manner in the following chapter, I will tackle the issue of what is meant by a 'constitutive relationship' in the second section of this chapter. Various notions of constitution are discussed in different fields of philosophy. I will go through some of them and ask whether they are applicable to the case of action and perception. I will conclude that we should adopt the approach that Tyler Burge (2010) takes on this matter, arguing that constitutive conditions are those conditions that determine the nature of something. Given this understanding, the claim that action is a constitutive condition on spatial perception amounts to nothing less than arguing that action is essential for spatial perception, that action is part of the nature of spatial perception. I think that this understanding of what is argued for in the accounts that I will discuss is most appropriate.

2.1.1 Action as a constitutive condition on spatial perception: Brief historical overview

The basic idea that spatial perception constitutively depends on action has quite a long and multifaceted history in philosophy and psychology. Many thinkers and schools have proposed it in the past. In this section, I will give a rough historical overview of the most prominent and important accounts that proposes such a relation. Of course, I cannot review or even mention here any account that argues in favour of action as a constitutive condition on perception. Hence, I present those accounts from roughly the last 150 years that I take to be most elaborate and/or influential on subsequent discussions.

The basic idea that perception of space cannot be explained by bare visual or tactile sensations came up at least in the middle of the 19th century. Alexander Bain (1855) was one of the first who proposed that a sense of movement was necessary for the perception of space.[31] Partly through the work of Carl Stumpf, Edmund Husserl got to know about Bain's approach.[32] Husserl's own stance on the question of how the spatiality of perception is to be explained always circled around this basic idea, esp. around the concept of *kinaesthesia*[33]. However, his interpretation of this notion changed fundamentally during his life. In his earlier works he took the concept literally and accordingly thought that sensations of movement are partly constitutive of spatial perception.[34] Later on, he revised his understanding to the effect that

31 His theory was criticized by a student of Brentano and Lotze (whose own students Wertheimer, Köhler and Koffka later became founders of the Gestalttheorie), the German philosopher and psychologist Carl Stumpf in Stumpf (1873), pp. 36–71.

32 Husserl read and made extensive notes on Stumpf (1873). He also possessed some later works of Bain; cf. Ferencz-Flatz (2014), p. 23.

33 The term '*kinaesthesia*' literally means sensation of movement. It goes back to the Greek terms *kinesis* and *aisthesis*, which mean movement and sensation, respectively. It was introduced not before the beginning of the 19th century.

34 In this time he discussed the questions as part of a Philosophy of Mathematics, namely as part of a philosophical account on geometry. This perspective on the problem is similar to the one of the mathematician Henri Poincaré who also expressed a tight relation between action and space in his writings, although without going into any detail. E.g., "It is

movement itself served the constitutive function. However, even this was not his final position. In his last writings and notes on this topic, *kinaesthesia* became more and more a synonym for the practical possibilities of the organism to move itself (see also Ch. 2.2.2.2 and 2.3.1).[35] Other prominent representatives of Phenomenology defended this view, especially in versions akin to the last Husserlian view. Heidegger thought that perceptual space is fundamentally structured by *care* (Sorge), which he took as a fundamental constituent of the way of being of humans, namely their mode of being as being-in-the-world (In-der-Welt-sein).

> But Dasein is "in" the world in the sense of a familiar and heedful dealing with the beings encountered within the world. Thus when spatiality is attributed to it in some way, this is possible only on the basis of this being-in. But the spatiality of being-in shows the character of *de-distancing* and *directionality*. (Heidegger 2010, p. 102)[36]

A contemporary of Heidegger's who is not much discussed these days also defended this kind of view. Max Scheler claimed that the grasped potentiality to move oneself constitutes perceptual space (Scheler 1928, esp. Part III; Scheler 1960, Part V). For him, drives and the dispositions to move they

---

this multiplicity of parries, and the resulting co-ordination, that is space" (Poincaré 2010, p. 104).

35 The first thoughts on space and spatial perception of his are developed in the 80's and 90's of the 19th century, when he planned to write a book on space that never was finished. These are published in Husserl (1983). The next important stage in his thinking about these issues is represented in his lectures from 1907 on "Thing and Space" (Husserl 1991). In his late writings, like Husserl (1954), questions about the constitution of space are treated, too (cf. about this last work Neumann 1999, pp. 132–8.) However, beyond this, there are many notes on the problem of constitution of space, several of which have not been published yet. Some can be found in Husserl (2008b), a larger piece in Husserl (1940b) and Husserl (1940a). See also the many references that are made in Claesges (1964), where almost only thoughts that are hided in the manuscripts (some of which are not published yet) are explicated.

36 "Das Dasein aber ist »in« der Welt im Sinne des besorgend-vertrauten Umgangs mit dem innerweltlich begegnenden Seienden. Wenn ihm sonach in irgendeiner Weise Räumlichkeit zukommt, dann ist das nur möglich auf dem Grunde dieses In-Seins. Dessen Räumlichkeit aber zeigt die Charaktere der Ent-fernung und Ausrichtung" (Heidegger 1972, pp. 104–5); cf. §§23–4 and Heidegger (1979), §§23–25, esp. §25.

ground are the decisive constituents of perceptual space (see also Ch. 2.6.2). Scheler, Heidegger and Husserl all emphasized in one or the other way the role of the practical engagement with the environment, the (dispositions to) bodily movements this leads to, the (potential) movements of the body. Other phenomenologists, like Maurice Merleau-Ponty, tried to express the same thoughts by investigating the structure and phenomenology of the Leib, which they took as a system of practical abilities and distinguished it sharply from a mere (physical) body (Merleau-Ponty 1974; see also Ch. 2.2.4.2).[37] In these conceptions, perceptual space is constituted by the mode of human existence as being an embodied existence. We find thoughts in this vein also in the current literature on embodiment, perhaps most prominently in the writings of Shaun Gallagher (2005; 2003). In his account, the physical structure of the body is most important while the practical dimension of embodied existence seems to be more in the background as is the difference between Leib and body (see also Ch. 2.2.4).

The development of the basic idea that started rather in the field of psychology than philosophy (although there was no sharp distinction between the fields then yet) was further developed in psychology as well. A cornerstone in this development is the work of Jakob von Uexküll, a prominent representative of ecology, who introduced the concept of 'Umwelt' (Uexküll and Kriszat 1956; Uexküll 1973).[38] His conviction was that the structure and possible contents of perception depend on the way of existence, the needs and practical possibilities the organism has. Thus, every kind of organism inhabits and perceives its own special Umwelt, its own ecological niche, as it were. This idea of species-specific ways of being and perceiving is also a key point in the theory of another important psychologist, James J. Gibson (1950; 1979). Gibson did not only propose that animals perceive affordances,[39] but he also claimed that perception depends fundamentally on the movements of

37 A similar conception is presented in Kruse (1974).

38 Central to ecology as a scientific discipline is the investigation of the relationships of organic systems with their environment, including mutual relationships of organic systems.

39 Gibson was not the first to use this concept. For example, Lewin (1982), p. 99 uses the term "Aufforderungscharakter" (this work was first published in 1933).

the organism, be it eye- or bodily movements (see Ch. 2.2.2.2). It was Gibson's ecological approach that was received by analytic philosophers such as Gareth Evans, whereby the basic thought that perception fundamentally depends on the activity of the organism found its way into analytic philosophy (Evans 1982; 1985).[40] Evans proposed that perceptual space depends on complex connections between sensory input and behavioural output (see Ch. 2.3.3).[41] Since this major figure of analytic philosophy, many philosophers have formulated similar views. A recent key account I want to mention here is Susan Hurley's. It presents another pronounced explication of the basic thesis in the Evansian tradition that is based on the distinction between instrumental and non-instrumental relations between action and perception (Hurley 1998; Hurley 2001; see Ch. 2.2.2).[42] Another major recent account in the field is the "enactive view" as elaborated by Alva Noë (2004). In this account, the explanation is mainly driven by sensorimotor knowledge, a kind of practical knowledge that the subject needs to possess and make use of to be able to perceive (see Ch. 2.2.3).

40 That Evans was familiar with the work of Gibson is shown by his explicit reference in Evans (1982), p. 122, n. 4. Evans also refers to Taylor (1978), an essay that explicates ideas of Merleau-Ponty's, and to Pitcher (1971).

41 Thus, he claims that "an egocentric space can exist only for an animal in which a complex network of connections exists between perceptual input and behavioural output." (Evans 1982, p. 154)

42 Other recent works that are worth mentioning here are Baldwin (1995; 2003), Malpas (1999), Grush (2000; 2007) and Schellenberg (2007).

Figure 2 is a chart of the different accounts that is sorted with regard to which aspect of action is taken to be crucial in constituting perceptual space. This rough overview, in which of course only the most important representatives of the basic idea are mentioned, shall show the great variety of thinkers, schools and accounts that express the thought that perception is fundamentally dependent on some activity on the side of the perceiver. In the next chapter, I will systematically contrast these different approaches according to their understanding of the central notions. First, however, it is necessary to take a closer look at the very idea of constitutive dependence: what it means and how it contrasts with related notions. This will be the task for the following section.

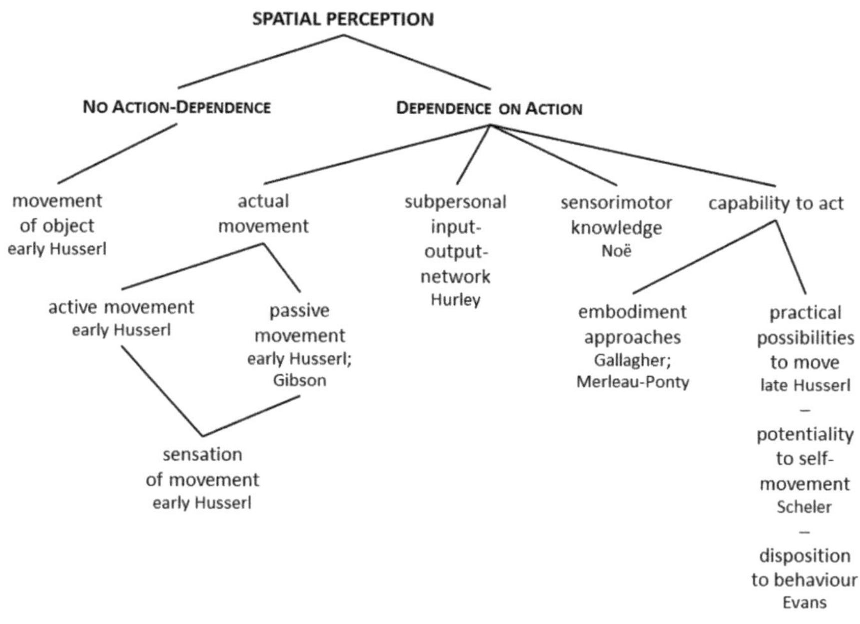

Figure 2: Overview of the different kinds of accounts that are discussed in the text.

2.1.2 The notion of 'constitutive condition'

Although there are a plethora of accounts that defend a constitutive relationship between action and perception, it is seldom explicated what such a relationship really means, which is quite surprising. After all, if one posits a special relation between two things one should be clear on what this relation amounts to and how it is to be conceived of. The following is not meant to be an exhaustive explication of the notion of constitutive condition. However, I want to get some grasp of the notion of constitution before evaluating accounts that claim that action is a constitutive condition on spatial perception.

I will proceed as follows. I will first shortly present different notions of constitution that do not seem applicable to the case of action and perception, although they include features that are also important for this case. These will be conceptions of constitution as a building relation, a conceptual relation, and as being distinct from causation, respectively. Neither seems to fit the case entirely. Therefore, I will concentrate on two accounts of constitutive explanation that are best applicable to the case of spatial perception and action. These will be the accounts of John McDowell, and Tyler Burge, respectively. I will argue that Burge's account is most convincing and I will show that his account incorporates features of all other discussed notions of constitution.

2.1.2.1 The various notions of 'constitution'

Constitution has been analysed as a kind of building relation (Bennett 2011), as a conceptual relation (Clark 1989), or by contrasting it with the notion of causation (Ylikoski 2013). Let me have a short look at each in turn. Although all three are not applicable *simpliciter* to the case of action and perception, they include features that help to understand the concept of constitution.

Karen Bennett (2011) analyses the relation of constitution as a "building relation". According to her, "a relation is a building relation iff it is asymmetric and irreflexive, and the 'input' relatum(a) is both more fundamental than the 'output', and in some sense overlaps it" (p. 92). This notion is best applicable to material objects, like a lump of clay constituting a statue. The first two properties (asymmetry and irreflexivity) indicate that the relation is a *directed* relation between two distinct entities. If x builds up y, then y does not build up x (asymmetry). In addition, x is not built up by itself (irreflexivity). The third property, (relative) fundamentality, indicates that the built entity is in some sense at a higher level than the building entity. Finally, the notion of overlap is necessary to mark that there is some connection between both entities.

I think this notion is not applicable to the case of action and perception. It is not clear why we should take the relation between action and perception as asymmetric. It seems possible that action will be constitutive of perception while perception is constitutive of action as well. It would be odd to rule out this possibility from the outset on *a priori* reasons. For the same reason, it is not clear why one should think of action being more fundamental than spatial perception. I think the only feature from Bennett's account that is important for the case of action and perception is the one that she tries to capture with her notion of overlap. I think this last property of Bennett's is meant to pinpoint to the notion of relevance: when some phenomenon is explained by appealing to another, the latter should be relevant for the former. However, there is a basic problem with her way of cashing out the notion: she seems to confuse the level of explanation and the level of things in the world. The mere fact that two things overlap does not tell whether one is relevant for explaining the other. This confusion of levels seems to be the reason why she ends up with a very unsatisfying idea of 'overlap' taken as a primitive notion that cannot be further explicated (pp. 91–2). I will come back to the topic of relevance below.

The second notion of constitution takes it as a conceptual relation. Clark (1989) argues, "fact *p* is constitutively relevant to *q* if on conceptual grounds we can see that *q* could not survive the subtraction of *p*" (p. 55, quoted from Colombo 2013, p. 556). On a strong reading, this quote implies that constitutive conditions are revealed by armchair reasoning alone. However, it is dubious why one should confine oneself to this sort of reasoning in investigating the constitutive conditions of spatial perception. I do not think that this is the right way to go. To the contrary, I think that empirical findings seem to be highly relevant for deciding this question (cf. Burge 2010; Merleau-Ponty 2002) and I will present a lot of relevant material in Part 3 of this dissertation.

In addition, the quote from Clark suggests that action is not more than a necessary condition on spatial perception. However, intuitively, a constitutive claim implies more than that. Otherwise, even the presence of oxygen would count as constitutive since without that, spatial perception is not possible. We want to say more than that action is necessary for spatial perception. The question is what this "more" may consist in.

In delineating the notion of constitution from the notion of causation, Petri Ylikoski (2013) considers more features of constitutive relationships. His characterization of a constitutive relation is as follows: "System S has causal capacity k in circumstances T, due to S's components $s_1, ..., s_n$ and their organization O" (p. 281).[43] Put more informally, explaining what constitutes some capacity of a given system means to elucidate components of the system along with their organization. The crucial distinction to a causal relation is that the relation of constitution does not hold between two events but between what he terms "causal capacities", which he takes to include all sorts of dispositional notions like ability, power, disposition etc. (p. 279). However, this way of spelling out the idea seems to be quite similar to Bennett's building relation – something Ylikoski himself mentions (p. 283). After all, the explanatory work is done by parts of the organism, their properties and organization.[44] Hence, unless one takes it natural to speak of action as a part of the organism, Ylikoski's notion does not seem applicable to the case of action and perception. This notion of constitution does not accommodate the case of one capacity constituting another one, a relation where it does not make much sense to speak of a part-whole relationship.

43 This formulation is not only similar to Bennett's 'building relation' stated above but also to the idea of a 'property theory' put forward by Cummins (1984), p. 14–22, condensed in the statement: "Successful analysis yields an explanatory payoff when we come to see that something having the kinds of components specified, organized in the way specified, is bound to have the target property" (p. 17); cf. Cummins (2000), esp. pp. 122–4.

44 The important bit of the characterization is the expression 'due to'. One might argue that it hides more than it explains.

This quick survey of different concepts of constitution does not only reveal that there are great differences in what is taken to be distinctive of the relation of constitution. Furthermore, it shows that the notion is not primarily developed to explain the case of two capacities constituting each other. Hence, these rather general approaches to constitution, although they point to important features like necessity or relevance, which will appear again below, are not of much use for my discussion.

In the following sections, I will discuss two accounts that better fit my case. I will finally argue that Burge's view is best for understanding what it means that action is a constitutive condition on spatial perception.

2.1.2.2 McDowell: Constitutive conditions and enabling conditions

John McDowell develops a distinction between constitutive conditions and enabling conditions on the basis of a critical discussion of Daniel Dennett's "cognitive theory of consciousness" (McDowell 1994; Dennett 1978). The central question of the debate is how the contents of conscious perception are to be explained. On Dennett's view, those portions of subpersonal level information becomes conscious that may be articulated verbally.[45] Thus, the content of visual experience consists in what is processed from the retina through the various parts of the visual system to that part of the brain where linguistic output is processed.

McDowell, in contrast, argues that studying the sub-personal mechanisms that underlie the processing of visual information cannot help us understand what perception is. These causal processes present merely *enabling conditions* for the organism to get in touch with its environment. However, they cannot be part of the conditions that are constitutive of perception. One can see the divergent opinions by Dennett and McDowell clearly by regarding the different interpretations of the following statement: 'It is in virtue of the subpersonal information-processing machinery that transmits its results to a certain

45 This is at least McDowell's interpretation. I am not concerned here with the question whether McDowell's reproduction of Dennett's ideas is entirely correct. I just take this discussion as the background against which McDowell's distinction of enabling vs. constitutive conditions shall become obvious.

part thereof that the organism can be said to perceive its environment.' While Dennett would read this claim, in particular the "in virtue of" phrase, constitutively, McDowell would read it as expressing an enabling condition.

Unfortunately, McDowell does not provide any clear positive view about what constitutive conditions would amount to. It seems that at least three contrasts are in play here, which shall illuminate what constitutive conditions are.

The first contrast is between constitutive and causal conditions. I already elaborated on this in the last section. McDowell argues against Dennett that just alluding to causal mechanisms cannot provide a constitutive explanation.

The second important contrast that is in play in his considerations is between the personal and sub-personal level of explanation. McDowell's discussion might be understood as arguing in general against the project of explaining personal level phenomena by sub-personal level mechanisms. It seems that sub-personal mechanisms are the wrong entities to draw on in explaining personal-level phenomena constitutively (although they might be necessary as enabling conditions). This might be because these mechanisms are causal relations (cf. first contrast). Alternatively, it might be because these mechanisms are mechanisms of just parts of an organism: when describing sub-personal processes one is not concerned with the organism as such but only with some parts of it. Since perception is a capacity of the whole organism, an exclusive reliance on sub-personal level mechanisms might, *in principle*, be insufficient to account for the personal-level property.[46]

Third, it seems to me that the leading contrast McDowell has in mind is the one between 'What?'-questions and 'How?'-questions. For the case of perception this means that the question 'What is perception?' (Answer: constitutive conditions) has to be distinguished from the question 'How is perception realized?' (Answer: enabling conditions). The latter can be answered by citing causal processes and the like based on investigations carried out in the sciences. However, these cannot, in principle, answer the (philosophically more important) 'What?'-question on his view.

46 If this were McDowell's rationale, it would seem very weak: we often explain certain capacities of things by alluding to parts and processes within the examined object.

However, again it remains unclear what form an answer to that question might take. The only hint is his interpretation of perception as openness toward the world, meaning that perception allows organisms to get "in touch" (p. 201) with their surroundings. This answer differs from the answer to the 'How?'-question in various aspects: it concerns the organism as a whole, including its way of living (in the ecological sense), instead of just some mechanisms within its systems; and the answer is based on rather general considerations, which are at best "remotely empirical".

These points show that McDowell does not provide any clear positive characterisation of constitutive conditions. Various aspects in his discussion seem to point to some distinctive properties of constitutive conditions. However, since one has to figure these out *ex negativo*, it is hard to get a grip on his positive views. Hence, I think that all we can take from him is that constitutive conditions answer certain 'What?'-questions and that they include more than just causal mechanisms.

2.1.2.3 Burge: Constitution and essence

A better articulated account of the notion of constitutive explanation is developed by Tyler Burge (2010, esp. pp. xiii–xvi; pp. 3–6; pp. 57–9; pp. 532–4). There are two crucial aspects about his analysis of constitutive conditions. The first is that they include necessary and/or sufficient conditions. The second is that what is thereby explained is something's nature, or what something is.

Constitutive explanations feature those conditions that cannot be changed without a change in the phenomenon to be explained. Condensed in a statement:

> A constitutive question concerns necessary (or sufficient, or necessary and sufficient) conditions under which something is what it is. Such a question concerns conditions under which something has the nature that it has. (Burge 2010, p. 57)

Giving an answer to a constitutive question is thus to spell out what is essential about some phenomenon, meaning the conditions that cannot be changed without any direct effect on the explained phenomenon.

Although I did not spell out the idea of constitutive conditions in terms of conditions for something's nature yet, it turns out that Burge's considerations about constitutive conditions link to almost all aspects I discussed in this section before. He thinks that questions about constitutive conditions are in a way conceptual questions but he makes clear that answers to these questions are approached neither by concept analysis alone nor without any empirical warrant (p. 5). This means that in the endeavour to understand the constitutive conditions for a given phenomenon we seldom if ever merely analyse the concept that applies. We take into account general empirical observations and scientific findings.

He also claims that the conditions that are part of constitutive explanations need not be part of the phenomenon explained (p. 58). There is thus not necessarily some part-whole relation between the constituting and the constituted. This speaks against the notion of constitution as some kind of building relation in the above-discussed senses.

In addition, he spells out his idea of the constitutive question by way of contrasting 'What?'-questions and 'How?'-questions (p. 533). Constitutive questions are regarded as 'What?'-questions. This is a direct parallel to McDowell, who claims that only constitutive explanations can tell what something is as opposed to merely telling how something is possible, which can be done by citing enabling conditions (see above).

However, how can Burge's analysis be applied to the case of action and perception? Taken literally, the claim that action is a constitutive condition on perception would then mean that action is necessary or sufficient or necessary and sufficient for perception. Which is the right option in this case?

It cannot be sufficiency. No one is able to perceive merely because he is able to act. There surely have to be some means of receiving and processing external input, like receptors and transducers of externally caused information, with which the subject picks up something from the environment. Thus, perception has some passive component. So what should we then think of the constitutive claim? We saw that it is neither enough to conceive of constitutive conditions as merely necessary (as in the exclusively conceptual reading above) nor to conceive of them as merely sufficient.

The claim that action is a constitutive condition on perception can only be intelligible if it is regarded as follows: action is one of the necessary conditions that is, together with other necessary conditions, sufficient for spatial perception. Neither is it alone necessary nor sufficient, nor necessary and sufficient, but only together with other conditions.

However, even this characterization is not enough. This leads us to a further aspect that is emphasized by Burge, namely the question of relevance.[47] The question is whether action is on a par with such necessary conditions for perception as the fact that oxygen is around. We can distinguish here between background necessary conditions and relevant necessary conditions. For this distinction there seem to be no general rules for how it is to be drawn. It depends on the explanatory project: the explanatory target determines which conditions are mere background and which are relevant (pp. 58–9). If one wants to explain the capacity of perception of humans, there are a number of necessary conditions that are implied simply by the fact that what we want to explain is a *human* capacity. All conditions that are necessary for the possibility of living humans are *ipso facto* background conditions for the possibility of human perception. In addition, we would count as background conditions those that make up the 'normal circumstances', i.e. the 'physical setting' in which we ordinarily perceive. However, this is not a general rule: a scientist might be interested in the question of how the physical and chemical composition of the human body and its surroundings have to look like for the possibility to perceive, e.g. what makes reflection, transmission and reception of light possible. In that enterprise, what we would regard as background conditions would figure as relevant conditions (cf. the notion of a proportionality between causes and effects in causal explanations proposed by Yablo 1992).

In sum, the relevance of certain facts always pertains to the particular explanation that one is after. For explaining spatial perception, this implies that all conditions that are necessary to make human life possible, the basic laws of nature etc. are taken as (possibly necessary) background conditions. The

47 This is the feature that Bennett tries to capture with her notion of overlapping, as I explicated above.

question is whether action is a relevant necessary condition on perception that, together with other relevant conditions, like the presence of an input processing machinery, is sufficient for the capacity of human perception.

We can even be slightly more specific. Since the *explanandum* in our case is *spatial* perception, action is supposed to be a constitutive condition merely on the spatiality of perception. Action might not be necessary for perception but only for *spatial* perception. This allows us to regard other necessary conditions for perception as mere background conditions and ask whether explaining spatial perception necessarily has to take into account the action of the subject.

2.1.3 Summing up

I think the discussion of the different concepts of constitution has shown that Burge's account of constitutive explanation best fits the case of action and perception. One might put what is at stake here by saying that the basic claim to be considered is that action is an essential feature of spatial perception. The various accounts I have already presented above in the short historical survey and that I will present in the next chapter specify the details of this relationship between action and perception in different ways. They share the commitment to the idea that there is such a dependence of spatial perception on action.

To sum up, what do we mean by the claim that action is a constitutive condition on spatial perception? We do mean that spatial perception is only possible for organisms that are capable of acting. In addition, we do mean that this capability for action together with other conditions that are necessary for the possibility of perception forms the ability of spatial perception of the organism. Thus, the claim is that action is part of the nature of spatial perception.

In the next two chapters, I will investigate different kinds of accounts that claim such a constitutive dependence of spatial perception on action. In the course of doing so, I will take into account almost all of the approaches that I mentioned in the historical overview. It will be obvious that there are quite distinct ways of spelling out the basic idea of a constitutive connection between action and spatial perception. Some accounts will be ruled out during the discussion already. Others will be kept until Part 3 and then evaluated against empirical evidence. What will emerge in the upcoming chapters is a comprehensive picture about the ways in which a constitutive dependence of spatial perception on action has been argued for and an evaluation of the appropriateness of the accounts. This will provide the basis for a justified decision about how to spell out the constitutive dependence, which in turn will finally ground the argument that egocentric representation is a kind of self-representation.

2.2 Systematic overview of constitutive accounts I: Hurley, Noë, Embodiment approaches

2.2.1 Introduction

In Ch. 2.1, I argued that the relevant notion of constitutive explanation for the case of action and perception alludes to the concepts of nature and essence. Claiming that spatial perception constitutively depends on action means to conceive of action as partly constitutive of spatial perception. In addition, as I also made clear in the preceding chapter, this basic idea has been cashed out in a variety of ways over the last about 150 years. In the following two chapters, I will develop a systematic overview of the different ways a constitutive dependence of spatial perception on action has been defended. The main distinguishing element among the accounts will be the way in which action figures in these approaches: the alternatives range from actual movements to mere action possibilities, from sub-personal level output states to knowledge about the sensory consequences of actions. Please refer back to Figure 2 (p. 67) to get a quick overview of the possibilities.

I will begin the discussion with a rather recent account given by Susan Hurley. The central contrast that figures in her approach is between constitutive and instrumental dependence of perception on action. Based on this distinction, it is possible to rule out certain accounts as merely claiming an instrumental dependence of perception on action. In particular, I will argue that all accounts that centre on actual movements effectively propose only an instrumental dependence of spatial perception on action. I will discuss two such accounts, in particular Gibson's ecological view and the account of the early Husserl. Thus, the left-sided accounts presented in Figure 2 are ruled out.

The remaining accounts that will figure in the discussion are the following:

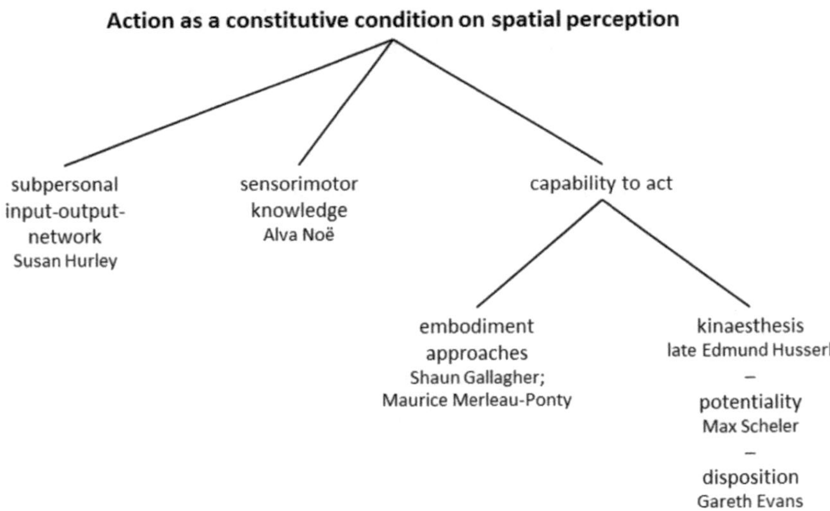

Figure 3: Remaining discussed constitutive accounts.

The chart shows that the remaining alternatives can be grouped into three main approaches. On the left, there is the approach developed by Susan Hurley that draws heavily on claims about the interconnections between input and output mechanisms. This approach contrasts with the others in that it draws on sub-personal level explanations and in that it is concerned with particular episodes of perception instead of the ability of spatial perception (see Ch. 2.2.2).

The second kind of approach is represented by the enactive account as put forward by Alva Noë. On his view, it is our basic knowledge about sensorimotor contingencies on which spatial perception depends constitutively. The account features in this overview since the relevant knowledge is taken to be practical knowledge of the effects of movement on sensation. A review of the position will elicit certain ambiguities related to the exact role that knowledge plays here. Thus, it is not entirely clear, and textual evidence is ambiguous on this point, whether spatial perception is an *exercise* of the practical knowledge or whether perception *depends* somehow on this knowledge. I will look at this ambiguity in detail in the respective section (see Ch. 2.2.3).

On the right side, there are several accounts that draw on the idea that the capability to act is crucial. I further divided this kind of approach into two sub-kinds, depending on the importance that is attributed to the body by the approaches. The accounts of Gallagher and Merleau-Ponty are developed in terms of preconscious interactions between body and world. However, it is difficult to figure out what really does the explaining there: mere physiology, body image or body schemas (see Ch. 2.2.4).

In Ch. 2.3, I will go on to discuss the accounts on the right side of the chart in Figure 3. These cash out the notion of capability to act in ways that are neutral with respect to embodiment (although they might be compatible with such approaches). Each of the approaches slightly differs in the way the capability to act is taken to be a constitutive condition of spatial perception. I will only briefly present the later Husserlian view, since his later thoughts on the issue are not systematically developed and difficult to interpret. I will mainly concentrate on the views of Scheler and Evans.

Overall, what emerges in the following two chapters is a comprehensive view about ways in which the basic idea of action as a constitutive condition on spatial perception may be developed. Each of the accounts centres on a different feature of action. Although there are some rather great differences among the approaches, it might not be impossible to integrate some of the accounts into another more comprehensive one. Hurley's two-level interdependence view, for example, might be combined with one of the other accounts that is formulated in terms of personal level notions only. Thus, I do not want to claim that all these approaches are mutually exclusive. However, this is also not the place for trying to integrate as much as possible into some 'master account'. I hope to make clear the basic tenets of each account and how it relates to the other ones. In Ch. 3.1, I will discuss some empirical evidence that will help to argue for one or another of the possibilities.

2.2.2 Susan Hurley's Two-level interdependence view

Susan Hurley's account of the dependence of spatial perception on action includes a wide range of philosophical and empirical considerations. Furthermore, in her discussion of the relationship between action and perception (Hurley 1998; 2001), many different questions are tackled in parallel, ranging from very general considerations about the overall structure of the mind and its relation to the world to quite specific questions about the relations among different mechanisms at the sub-personal level. Her project is ambitious: she argues not only against what she takes to be the traditional understanding of the relationship between action and perception, but also against the received understanding of the relationship between the personal level notions of perception and action and the sub-personal level mechanisms from which they emerge.

Two general features are distinct of Hurley's approach compared to the other accounts that I will discuss: It explicitly takes into account sub-personal level mechanisms, and her approach centres on particular episodes of spatial perception instead of the capability of spatial perception.

Hurley develops her account of the constitutive dependence of perception on action by contrasting it with another form of dependence that she calls instrumental dependence. Both are ways in which perception can be dependent on action. In claiming that perception can constitutively depend on action, she questions the traditional framework of how action and perception are connected. Within that traditional framework, only instrumental dependence is possible.

The contrast between instrumental and constitutive dependence will also help to see that various kinds of approaches mentioned in the historical overview in the last chapter are actually not proposing a constitutive dependence of perception on action. In particular, I will show that the accounts of the early Husserl and of Gibson merely defend an instrumental dependence of perception on action. The critique against these views is applicable to all accounts that argue that actual movements are constitutive of spatial perception.

I will first present Hurley's account of constitutive dependence based on the contrast to instrumental dependence. In the second step of this section, I will show that the accounts of Gibson and the early Husserl propose an instrumental dependence and not a constitutive one.

2.2.2.1 Instrumental dependence and constitutive dependence

One of Hurley's general purposes in developing her account is to argue against what she terms the 'traditional view' about action and perception. On this view, action is taken as output from mind to world and perception is seen as input from world to mind.[48] This picture includes a distinction between personal level notions, like experience and action, and sub-personal level notions, like sensory inputs and motor outputs that are conceived of as causal processes (Hurley 1998, p. 213).[49] It implies a deep separation between perception and action: perception is exclusively determined by the causal processes on the input side, while action is exclusively determined by the causal processes on the output side. Hurley's general agenda is to argue against this strong division. She aims to show that action and perception are determined by complex and dynamic interactions among sub-personal input and output processes – leading to a corresponding interdependence on the personal level, between perception and action.

48 She dubs this model the 'Classical Sandwich': action and perception are the bread while the mind is the filling; cf. e.g. Hurley (1998), p. 401.

49 The distinction between the sub-personal level and the personal level of explanations was originally introduced in Dennett (1969).

Hurley's crucial insight is that we have to distinguish between instrumental and non-instrumental dependence between perception and action and that only the latter expresses a constitutive relationship. The distinction is made in terms of directness. Whereas in the case of instrumental dependence, action has an impact on perception only indirectly, the impact is direct in the non-instrumental dependence. I will first introduce the notion of instrumental dependence with an example. This will help to make clear what is distinctive about the contrastive case of constitutive dependence.

Suppose you are at a royal wedding ceremony, thousands of others are there and you want to get a glimpse of the bridal couple, which is not easy in the middle of all those people. You move your head, stand on tiptoes etc. to see the couple. In other words, you act in certain ways in order to get to view certain things. One can put this as a means-end relation where the actions you undertake are the means to accomplish the end, which consists in seeing the royal couple. In a sense, this case exemplifies a relation of dependence: what I perceive is determined by how I move. I could not see the royal couple if I would not move thus and so (*ceteris paribus*). The movements are a means to make possible a certain perceptual content. Therefore, the dependence is aptly called 'instrumental'.

The dependence is indirect: movements alter my perceptual contents by altering my perspective on the world; they alter the causal configuration between the things in the world and my sensory systems by altering their respective spatial relationship. What could not have a causal impact on me in advance of the movement has a causal impact due to the movement.

This is the general structure of the instrumental relation: a certain perception is brought about by means of movements that establish a causal relation between the subject and an object that did not obtain in advance of the movements.

These movements are not necessarily part of intentional actions as in the example where there is an intention to see the bridal couple that leads to certain movements that serve this end. Thus, no active exploration of the environment is necessary for the instrumental case. The basic structure of instrumental dependence is present even if movements are non-intentionally made and/or go on completely unnoticed (think of saccadic eye movements). In all these cases, movements change the configuration of the pattern of light on the retina.[50]

Furthermore, a characteristic of the instrumental relation is that passive movements, in principle, could accomplish the same effects on perceptual content. To change the pattern of relevant stimuli it is not necessary that I move myself. Nothing speaks against the possibility that I am moved with the same result (think, e.g., of a king that is carried by his adjutants and orders them to lift him a little bit up so that he can see something behind a wall).

Finally, even movements of the environment can have the same effect in principle. This is because the core of the instrumental dependence consists of a changing causal relation between subject and world. Any sort of movement can bring about a change in this relation, be it of the subject or the environment.

On Hurley's account, every relationship between perception and action is instrumental where perception changes because of output states changing input states. She frames her account in terms of motor theories that emphasize the role of neural motor signals in explaining perception and draw on notions like efference, afference, and efference copy (cf. Gallistel 1980, Ch. 7; Hurley 1998, pp. 436–7). The notion of 'reafference' can be taken as a prime example of instrumental dependence. The more general concept of 'afference' is another word for input, referring to any information that proceeds from the periphery toward the central nervous system; the opposite notion is efference, or output, e.g. motor signals (Karnath and Thier 2012, p. 798). Reafferent signals comprise those afferent signals that are caused by the subject herself, thus being a kind of feedback. For example, in moving one's limbs one receives input that is caused by these movements itself; when quickly turning around in an otherwise static environment, a certain visual flow is generated

50 I will confine myself here to cases of visual perception, since these are the most discussed ones and usually serve as the paradigmatic case. This does not imply that for other sensory modalities analogue considerations do not apply. For example, you fulfil certain movements in order to be able to touch something.

that is caused solely by this movement. These are examples of reafferent signals, viz. input that is generated by movements, or output of the subject.

Perception depends instrumentally on action in case a change in perception is brought about by a change in input elicited by movements of the subject. This is the structure of instrumental dependence. Let me now turn to Hurley's notion of constitutive dependence of perception on action.

What are the characteristics of the constitutive (non-instrumental) dependence? The core idea is that action has a *direct* impact on perception. What this 'direct' means has to be brought out by contrast to the 'indirect' dependence that is the core of the instrumental dependence.

As we saw, an instrumental, or *indirect*, dependence of perception on action is instantiated when a change in the causal relationship between the environment and the sensory systems of the organism leads to a change in perception. In this case, action is a means to change the input pattern of what reaches e.g. the retina.

In contrast, a *direct* impact of action on perception does not proceed via this classical action perception loop. Thus, the constitutive dependence comprises cases where input is constant while there being an effect of output on perception. A prototype of such a non-instrumental dependence is the effect that 'efference copies' have on perception. Efference copies (or 'corollary discharge') are signals that are a copy of actual output signals and that are transmitted to other brain regions, providing a kind of instant feedback about how the system is about to behave. These signals are also transmitted to brain areas that compute perception-related processes. Given this, the following situation is intelligible as an instance of non-instrumental dependence: there is a constant input pattern, a changing output pattern (efference copy), and a changing perception. Only the output pattern and perception change, input does not. Therefore, a non-instrumental dependence is instantiated (Hurley 1998, p. 436).

A vivid clinical example given by Hurley are cases of paralyzed eye (pp. 372–3; cf. Gallistel 1980, p. 175). Patients with a paralyzed eye are still able *to try* to move their eye. Neural signals are transmitted that otherwise would lead to a corresponding eye movement. The surprising fact about these attempts to move the eye is that, although the retinal pattern is constant, they have an effect on perception. The world seems to move. Clearly, this illusionary impression does not result from changes of input (as in the instrumental case) since the eye cannot move. The only thing that changes and therefore has to be taken into account to explain the phenomenon is the attempt to move the eye.[51]

The essential features of Hurley's notion of constitutive dependence can be summarized as follows. The core contrast between constitutive and instrumental dependence boils down to the contrast between direct and indirect impact that output can have on perception. Whereas in the instrumental case output determines perception by way of altering the input patterns on which perception relies, in the constitutive case there is a direct impact of action on perception. Thus, in the latter case output becomes part of the basis on which perception relies. "In distinguishing instrumental from non-instrumental dependence, we're trying, roughly, to capture the difference between action having an effect on a separate perceptual system and its sharing the constitution of that system" (Hurley 1998, p. 363).

Instrumental and non-instrumental dependence can both be cashed out in terms of a causal relation. This is especially clear for the instrumental case, but Hurley explicitly states that also the non-instrumental dependence can be a causal relation: "It is important to note that non-instrumental dependence here can be both a constitutive and a causal relation [...]" (Hurley 2001, p. 15).

Finally, the contrast between instrumental and non-instrumental dependence is made in terms of particular episodes of action and perception. This distinguishes Hurley's account from the other accounts I will discuss.

Let me add a few general notes about Hurley's approach. Her view is quite liberal in at least two respects: First, instrumental and constitutive dependence are not thought to be mutually exclusive. Both of them contribute in important ways to the way we perceive the world. Furthermore, she neither claims that every case of perception nor every aspect of perception is constitutively dependent on action, or output. She opts for a pluralistic view, in which different relationships are needed to explain different cases. For some

51 For more examples, see Hurley (1998), pp. 371–9.

cases, even the traditional view presents an intelligible explanation. In general, Hurley's strategy to challenge the traditional view about perception and action is to argue via counterexamples from neurology and via thought experiments and then generalize the structures found there. She is aware that what she gets with this strategy is only a refutation of the traditional view as a general conceptual framework for how perception, action and cognition work together, which does not rule out that for some explanatory enterprises the traditional view might be appropriate. She regards it as an empirical matter (Hurley 1998, p. 340). In effect, her view combines personal level approaches like the ecological approach (which she appreciates although it merely posits instrumental dependencies) with sub-personal level approaches like the motor theory of perception. Perception constitutively depends on action in virtue of the complex and dynamic sub-personal mechanisms among input and output, or afference and efference (Hurley 1998, p. 438).

I will discuss some more details of her account in Ch. 3.1.2. In the second section of this chapter, I will argue that the accounts of Gibson and the early Husserl fall short of arguing for a constitutive dependence of perception on action. Rather, both deploy an instrumental dependence in Hurley's sense.

2.2.2.2 Applying the notion of instrumental dependence to the accounts of Husserl and Gibson

To recap, the instrumental dependence essentially consists in a changing spatial *relation* between subject and environment that gives rise to certain perceptions. A change in this relation can be brought about by changes in either of the *relata*. Therefore, it does not rely on self-movement (active or passive) *necessarily* and hence it does not hold necessarily between *action* and perception but only between *movement* and perception.

Two of the accounts we mentioned in the historical overview are of this sort: the early Husserlian account and Gibson's ecological account.

Early Husserl

The instrumental dependence is present in the early writings of Edmund Husserl (Husserl 1983; 1991). In the historical overview in Ch. 2.1.1, I already noticed that the key concept in Husserl's writings on the subject is *kinaesthesia* (cf. Claesges 1964, p. 64, n. 1). Literally speaking, the term means 'sensation of movement'. This was the meaning Husserl had in mind in his earliest writings on the issue.[52] With regard to visual space perception, he claimed then that the association of visual sensation and sensation of movement grounds three-dimensional space, whereas vision alone only constitutes a two-dimensional visual field.[53] The central idea is thus that visual perceptual space is constituted by an association between visual sensations and sensations of movement.

In the first writings on perceptual space from the early 1890's, Husserl's ideas are still very vague and unsystematic. But at least two convictions are already present that remain important for him: on the one hand the general primacy of the psychological investigation, i.e. the descriptive and genetic analysis of perceptual representations (Husserl 1983, pp. 302–3) and, of course, the idea that movement has to play an essential part in that analysis. This is further developed in his lecture "Ding und Raum" ("Thing and Space") (Husserl 1991) from 1907, which is the first systematic account from Husserl on this question. In this lecture the meaning of *kinaesthesia* sometimes shifts to movement, sometimes stays with sensation of movement (cf. Ferencz-

52 In large parts of the literature on Husserl's account of perception, researchers (in my view mistakenly) argue that Husserl meant something like movement itself or even the system of possible movements. Cf. e.g. Claesges (1964), p. 64, n. 2: "Die Kinästhese kann in ontologischer Einstellung als ‚Organbewegung' angesehen werden." ("In ontological perspective, kinaesthesia can be regarded as 'movement of an organ'"; my translation.) Cf. Neumann (1999), p. 137; Drummond (1979), p. 22; Mattens (2010), p. 171. As already said, the term 'kinaesthesia' was subject to various substantial revisions during Husserl's life. The significant shift in meaning of this central term is well described in Ferencz-Flatz (2014).

53 Husserl (1983), p. 305: "In der Natur der Tast- und Bewegungsempfindung ist der dreidimensionale Raum des Blindgeborenen begründet. In der Natur der [sic!] Gesichtssinnes ist begründet das zweidimensionale Gesichtsfeld. Dazu die besonderen Lok<al>zeichen und endlich die Verbindung der Gesichtsempfindung mit Bewegungsempfindung, wodurch eben immer nur ein dreidimensionaler Gesichtsraum erwachsen kann." ("Three-dimensional space for the born-blind is grounded in the nature of sensations of touch and movement. The two-dimensional visual field is grounded in the nature of vision. A three-dimensional space can only arise from local signs and the connection between visual sensation and sensation of movement"; my translation.)

Flatz 2014, pp. 24–8). Nevertheless, it is important to note that on this account a succession of particular movements constitute perceptual space. "Alle Räumlichkeit konstituiert sich, kommt zur Gegebenheit, in der Bewegung, in der Bewegung des Objektes selbst und in der Bewegung des ‚Ich', mit dem dadurch gegebenen Wechsel der Orientierung" (Husserl 1991, p. 154; "All spatiality is constituted, comes into reality, by movement, by movement of the object itself and by movement of the 'I', by the thereby given change in orientation"; my translation). The perception of a physical body is thus constituted by a succession of percepts. "Der Körper muß sich drehen und verschieben, oder ich muß mich, meine Augen, meinen Leib bewegen, um ihn herumgehen und mich zugleich nähern und entfernen; oder endlich, beides muß stattfinden" (pp. 155–6; "The object has to rotate and displace or else I have to move myself, my eyes, my body [Leib], go around it and at the same time approach it and diverge from it; or, finally, both has to happen"; my translation).

The basic idea seems to be the following: To perceive a physical and hence spatial object, it is necessary that certain movements happen to the effect that different sides of the object are presented to the subject. The presentation of the different sides has to happen in continuous movements to ensure that it is the same object to which the different sides belong. The movement can happen on the side of the subject or the object. The movements make possible the presentation of the object (p. 169). Thus, the movements (or the correlated sensations, respectively) are not *essentially* connected to visual sensation but only *functionally* (p. 170).

This clearly is a case of an instrumental relation between movement and spatial perception. All three elements of this kind of relation, as stated above, are in place: it is just movement (not action) that serves the function of constituting spatial perception; and it is only the relationship between subject and object that is crucial, hence, it is not even necessary that the movement happens on the side of the subject, movement of the object is enough.[54]

54 Husserl's conception of the constitution of perceptual space is, of course, much more complex than stated here. He distinguishes different kinds of kinaesthetic systems, like

The early Husserlian view thus does not express a constitutive dependence of spatial perception on action. Action is merely taken as a succession of movements that change the input in a way to allow for the perception of spatial properties. This is thus an archetype of an instrumental dependence of perception on action.

The second approach that rests on a similar claim is Gibson's ecological view of spatial perception to which I want to turn now.

Gibson's ecological view

The second historical example of a merely instrumental relation between spatial perception and action can be found in the account of Gibson (1979). For Gibson, perception is a process that is undertaken and explained by the activity of the whole perceiving organism (and not just by processes on the retina and in the visual cortices). It is the process whereby the "eye-head-brain-body system registers the invariants in the structures of ambient light" (p. 61). This registration fundamentally depends on movement: Gibson distinguishes two 'layers' of the ambient optic array, a "nonessential" layer that is constantly changing with movement, and an "essential" layer that stays invariant despite change (p. 73).

His description seems to imply a hierarchical picture of how the optic array is structured; however, I take it that his account is better described as proposing that the optic array has different kinds of elements or aspects, namely, variant and invariant (p. 74). The crucial claim is that the latter are only revealed by change in the former. "What is invariant does not emerge unequivocally except with a flux. The essentials become evident in the context of changing nonessentials" (p. 73).

Hence, perception consists in registering invariants; invariants are only registered when there is change in the array of light that hits the eye; therefore,

---

the possible set of eye movements or the set of head movements that are necessary for different aspects of perceptual space. In general, the idea seems to be that the set of the dimensions of possible movements determines the set of dimensions of the associated perceptual field. Thus, the two-dimensional space of movement of the eye is correlated with the two-dimensional visual field. A detailed discussion of Husserl's account in "Ding und Raum" can be found in Drummond (1979).

perception is tight to movement: "Observation implies movement, that is, locomotion with reference to the rigid environment, because all observers are animals and all animals are mobile. Plants do not observe but animals do, and plants do not move about but animals do" (p. 72). Gibson surely thinks that movement is a necessary condition on the possibility of perception. In addition, he thinks that bodily acts of the perceiving organism are the most important means to register the relevant invariant structures from the optic array.

However, his approach nevertheless posits only an instrumental connection between action and spatial perception. There is nothing in the theory that is specifically tied to *action*. All that is necessary is the generation of a visual flow by which invariants can be detected. Therefore, although Gibson often speaks about movements of the observer being crucial, other means of inducing change to the "superficial structure" of the optic array can have the same effects, too. As sources of change in this structure, he explicitly counts events in the environment, movements of the subject, and changing lighting conditions (p. 87).

Thus, Gibson's approach clearly endorses the characteristics of the instrumental relation of action to spatial perception. What is crucial is only movement, whether active or not does not make any difference; even movement in the environment is sufficient since effectively it is a changing relation between subject and environment that is explicatory (cf. Mossio and Taraborelli 2008, p. 1332).

Accounts like that of the earlier Husserl and Gibson that claim that successions of sensation brought about by movements explain spatial perceptual space state only an instrumental relation, in which movement is a means for the organism to perceive spatial features. Therefore, it is no surprise that the explanatory weight in these conceptions is not carried by anything that is essentially related to action but only by something that is contingently related to (some kinds of) action, namely movement.

We can see thus that, in general, approaches that centre on actual movements in cashing out the dependence of spatial perception on action fall short of arguing for a constitutive dependence. Any account that relies on the idea that spatial perception is dependent on a changing spatial relation between subject and object defends a mere instrumental relation (see figure 4).

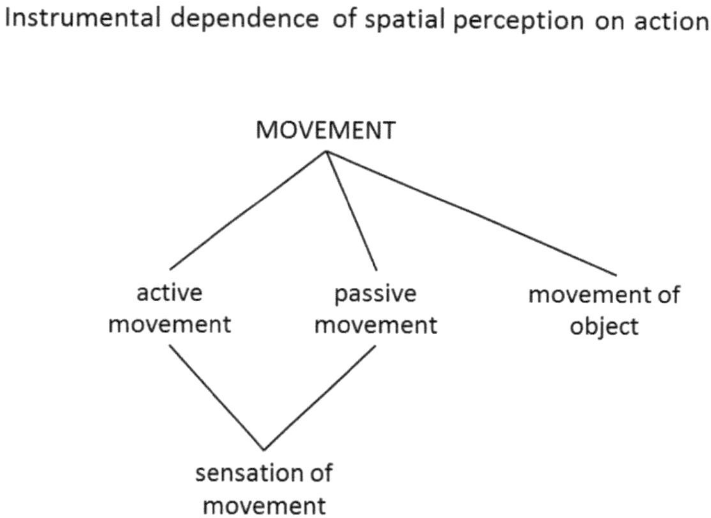

Figure 4: Proposed understandings of action that make up a mere instrumental relation between action and spatial perception. Action in this case is understood as actual passive or active movement of the subject. A mark of the instrumental dependence is that, in principle, the same effect can be obtained from movement of the subject, or of the perceived object. A further possibility is that not movement but sensations of movement on the side of the subject are crucial.

2.2.3 Alva Noë's enactive approach

On Alva Noë's (2004) view, what constitutes spatial perceptual content is a certain kind of practical knowledge. The relevant knowledge is not a form of propositional knowledge (pp. 117–22). The content of the knowledge in question consists in sensorimotor contingencies: it is a grasp of the effects that movement *would have* on sensation. In most parts of his work, this idea is spelled out with reference to examples of perceiving particular objects. For example, when one sees a tomato one always sees only a certain side of it. Nevertheless, it is compelling to say that the content of one's perception is not just specified as "part of a tomato" but "a tomato" (taken as a solid, voluminous, three-dimensional object). How is it possible that we can attribute to ourselves a perception of an object when all that we see is just one particular side of it?[55] Noë's solution to this problem, which he calls the 'problem of perceptual presence' (pp. 59–65; p. 77), is that we see the tomato since we "understand, implicitly, that [our] sensory relation to those parts [that we do not, strictly speaking, see] is mediated by familiar patterns of sensorimotor dependence" (p. 77). In other words, we see the tomato as something with certain spatial properties like size and shape thanks to our implicit knowledge of how we would see the currently hidden parts if we, or the tomato, would move thus and so. Here is his view condensed in one statement:

> We are not given the visual world all at once as in a picture; we must reach out and grasp the detail (as it were) by movements of our eyes and head. We possess the sensorimotor knowledge to be effective in our exploration. It is this mastery that is the basis of our sense of the presence to vision of what is in fact beyond our reach. It is this mastery that is the basis of perceptual content. (p. 75)

The same way of explanation also pertains to the question of why we see objects as left, right, up or down. In this regard, Noë argues: "To experience an object as off the left is to experience it as standing in a relation to one which one grasps as constituted by patterns of sensorimotor dependence" (p. 87). Again, what constitutes the spatial content of my perception is an implicit understanding of the changes to my sensation elicited by certain movements. In this regard, he also talks about "sensorimotor profiles". These pertain either to (kinds of) objects (p. 78) or to positions in egocentric space

55 This phenomenon was also tackled by Husserl under the heading of the "horizontal structure" of experience on various occasions; e.g. Husserl (1991), pp. 55–60. Noë seems unaware of this.

(p. 87), specifying how sensory stimulation varies with movement. The sensorimotor profile is thus the sum of the known sensorimotor contingencies related to an object, a kind of object[56] or a point in egocentric space.

In sum, on Noë's account, one can only have spatial perceptual contents if one has a practical grasp of how one's sensation would alter if one were to move in relation to the seen object or scene, that is, if one grasps the sensorimotor profile of what one sees. In a modification of Kant's famous statement that intuitions without concepts are blind, Noë says, "intuitions [...] without knowledge of sensorimotor significance of those intuitions, are blind" (p. 11).

Let me now I turn toward the role that movement plays in Noë's account. There are three possible reasons to suppose that on his account spatial perception is only instrumentally related to action in terms of Hurley's notion, which I introduced in the last section. The first reason is that actual movements are necessary for every case of spatial perception, which some statements made by Noë suggest. The second is the observation that at some point in the development of the relevant sensorimotor knowledge actual movements are indeed necessary, namely in the initial stage when the knowledge is learned by interaction with the environment. The third reason concerns the fact that Noë's account at least relies on counterfactual movements. I will discuss these reasons in turn.

The first reason for regarding the account of Noë as proposing merely an instrumental dependence between action and spatial perception consists in his statements that indicate that vision is similar to touch (pp. 16–7), i.e. an active exploration of the environment: "[P]erceiving is an activity of exploring the environment, one that requires the perceiver's mastery and exercise of sensorimotor knowledge [...]" (Noë 2008, p. 694).[57] Formulations like this

56 It is not entirely clear whether the sensorimotor profile pertains to particulars or to kinds. I think it is plausible to suggest that the profile pertains to kinds of objects. All spherical objects, e.g., share the same sensorimotor contingencies.

57 Cf. Noë (2004), p. 75: "We are not given the visual world all at once as in a picture; we must reach out and grasp the detail (as it were) by movements of our eyes and head. We

suggest that perception depends on actual movements, that one has to move in relation to one's environment for there to be any spatial perception at all, while the role of sensorimotor knowledge is to guide these explorative movements. If true, this would obviously render the account into one that merely states an instrumental relation between perception and action, similar to the accounts of Husserl and Gibson that I discussed in the preceding chapter. Since on such a view, perception is dependent on movement in that movements change the causal setting of subject and environment thereby making possible perceptual content. However, I think one should not take these statements at face value. Other statements indicate that he does not think that perception depends on actual movements. Thus, I think these statements are rather careless formulations of his view that should not base the interpretation of his account. I want to present an alternative interpretation of his account that does not imply this consequence. Thus, the first of the three reasons to think that Noë effectively proposes only an instrumental dependence of perception on action is not convincing.

The second reason is that the development of sensorimotor knowledge depends on the exploitation of the instrumental dependence between action and perception. Thus, at least in the aetiology of the relevant sensorimotor knowledge, actual movements play a crucial role. The knowledge has to be acquired somehow in the first place. One might suppose an explanation of the development of this knowledge that draws on phylogenetic developments, which would rule out the threat of a mere instrumental dependence. However, the following three observations indicate that that Noë has an ontogenetic development of sensorimotor knowledge in mind.

First, he repeatedly refers to results from studies with inverted goggles. In the studies he refers to,[58] participants report that their perception is initially

possess the sensorimotor knowledge to be effective in our exploration"; p. 228: "Perceptual experience, according to the enactive view, is an activity of exploring the environment drawing on knowledge of sensorimotor dependencies and thought."

58 He refers to Stratton (1897), Kohler (1951) and Taylor (1962); cf. Noë (2004), pp. 7–11, pp. 91–6.

inverted but restored after a while, and especially after they get used to interacting with their environment while wearing the goggles.[59] He argues that what accounts for the supposed changes in spatial layout of perception that subjects wearing such goggles undergo is the change in sensorimotor knowledge. Therefore, there seems to be some 'plasticity' of sensorimotor knowledge that would not be expected if it were the result of a phylogenetic development. I take it that properties of our ways of information processing in general that are to be explained by phylogenetic developments are generally such that it is very hard if not impossible to alter them.

A second reason for supposing an individual ontogenetic development is his reference to cases where congenitally blind people come to see due to surgery. In these cases, it takes some time until, if at all, normal vision is established. Noë interprets this as the acquiring of the relevant sensorimotor knowledge (Noë 2004, pp. 3–7).

Finally, one can find passages like the following two. "Only through *self*-movement can one *test* and so *learn* the relevant pattern of sensorimotor dependence" (p. 13).[60] "Perceptual experience acquires spatial content thanks to the establishment of links between movement and sensory stimulation" (p. 102).[61]

In sum, one can assume that the relevant sensorimotor knowledge is acquired individually by learning associations between movement and changes in sensation. Therefore, at least initially for the establishment of sensorimotor knowledge, he posits a mere instrumental relation between movement and spatial perception. The general idea seems to be that the exploitation of the instrumental relation makes possible the acquisition of the relevant

59 There are some issues regarding how to interpret the results of these studies. Furthermore, not all results could be reproduced; cf., e.g. Linden et al. (1999) and Smith (manuscript), pp. 7–9. See also section 3.1.3 above for a detailed discussion of these cases.

60 I cannot see that self-movement is necessary for the learning of sensorimotor contingencies on this approach. Movement is also experienced in case it is passive. Therefore, the relevant dependencies of changes in sensation to patterns of movement might also be learned by passive movements. Noë does not argue for his claim that active self-movement is necessary.

61 This is also strongly reminiscent of views held by the early Husserl, e.g. Husserl (1983), pp. 308–9.

knowledge, which then takes over the role of actual movements. Later on, one does not have to move in order to perceive because one already knows (or grasps) what would happen if one were to move.

One might argue that, given the relevant sensorimotor knowledge initially and necessarily relies on actual movements, the account ultimately states merely an instrumental relation between action and perception. Self-movement is the means to enable me to see spatial properties in that I unify a series of percepts that are generated by my movement into a persistent spatial perception. This is at least a necessary developmental stage according to the theory.

However, against this consideration one can argue that the fact that the relevant sensorimotor knowledge is established by exploiting the instrumental relationship between action and perception does not imply that the dependence of perception on this knowledge is itself instrumental. Hence, even if the instrumental relationship between action and perception does play a crucial role in the development of the constitutive relation between sensorimotor knowledge and spatial perception, it is not a convincing reason to regard the latter relation as merely instrumental as well.

Let me turn to the third reason one might have to argue that Noë only defends an instrumental dependence. One might interpret the account as claiming that particular counterfactual movements constitute spatial perception. On this interpretation, it would still be the case that what explains my perceptual content are actual movements and the changing causal relation they elicit, only that I do not have to actually conduct the movement because I already know of the consequences the movement would have for my sensation. What ultimately explains the perception of spatial properties is the changing causal relation between perceiver and environment and even so in mature humans since there, the process is present in the knowledge of how sensation would alter by certain movements one had acquired before. The practical knowledge can be taken as a mere surrogate of actual movements that serves the same function. The following quote from Noë fits nicely with these considerations: "It is intuitively plausible, for example, that the circularity of a plate becomes available in encounters with its changing perspectival shape as one moves (or would move) with respect to it [...]" (p. 228).

Does this line of thought ultimately render the proposed relation between action and perception into a mere instrumental one? Before, I argued that the core of the instrumental relation is that perception is dependent on action in the sense that action changes the causal relation between the perceptual systems and the environment, what enables the subject to perceive things it would not perceive without movement. Given this, the relation between sensorimotor knowledge and perception is not instrumental on Noë's account. There is no changing causal setting, since there is no actual movement happening. The relation of sensorimotor knowledge to perception can be taken as a constitutive relation in terms of Hurley's notion.

To sum up, how is the idea that action is a constitutive condition on perception cashed out in Noë's account? Action plays a crucial and necessary role in the establishment of sensorimotor knowledge, which plays the key part in this account. In the development of this knowledge it is the instrumental relation of action to perception that is exploited, it is the learning of how the sensational pattern of (kinds of) objects varies with patterns of movement. What is constitutively linked to perception is the sensorimotor knowledge itself that is established throughout this development. I have considered three reasons to suppose that Noë merely defends an instrumental dependence of perception on action. I think that all three considerations are inconclusive.

However, I will come back to Noë's approach in Ch. 3.1.4, where I will argue based on experimental findings using sensory substitution devices that the sensorimotor knowledge approach is unconvincing since there are cases the knowledge about how sensation would alter with movement does not lead to spatial perception. This speaks against the central claim endorsed by Noë that spatial perception consists in a mastery of this sensorimotor knowledge.

With these considerations about Noë's approach, I will turn to accounts that claim in one or the other way that dispositions to action form a constitutive condition on spatial perception. The final section of this chapter is devoted to two accounts that centre on the embodied nature of human subjects, namely those proposed by Shaun Gallagher and Maurice Merleau-Ponty.

2.2.4 Embodiment approaches: Shaun Gallagher and Maurice Merleau-Ponty

In this section, I want to discuss two accounts that assign a crucial role in the explanation of perceptual space to the body. These are the accounts of Shaun Gallagher and Maurice Merleau-Ponty, respectively. Both accounts figure in this discussion of the claim that action is a constitutive condition on spatial perception since in both accounts the practical bodily engagement with the world is taken as the *explanans*. In both accounts, the relevant notion of action is inextricably linked to embodiment. It will turn out that the notion of the body schema is central for both positions.

On both views, the practical engagement of the body with the world that is beyond the conscious access of subjects is supposed to explain the spatial structure of perceptual space. Thus, action figures in these accounts as the anonymous and habitual workings of bodily mechanisms, which ground fluent interactions with the world even without conscious control. Examples for this kind of practical involvement are such capacities as maintaining balance, quickly reacting to external stimuli without any planning or deliberation, grasping things in the vicinity without conscious control etc. All these activities can be fulfilled automatically, and indeed are often fulfilled this way in daily life. It is with these activities of the body that the notion of body schema is most strongly associated. We will see that in Merleau-Ponty's account the position finds its most radical articulation in the claim that the living body is a subject in its own right. He claims that the activities of this subject constitutes the nature of perception and experience that is phenomenally available to us as persons.

This section has two parts. First, I will present Gallagher's position, which is unfortunately quite ambiguous on central questions, as we will see. Merleau-Ponty's account, as articulated in his *Phenomenology of Perception*, presents a much more precise articulation of the same basic ideas. I will end this section with a brief summary of both accounts.

2.2.4.1 Embodied action I – Shaun Gallagher

In his monograph *How the Body Shapes the Mind*, Shaun Gallagher claims that "one of the important functions of the body in the context of perception and action is to provide the basis for an egocentric spatial frame of reference" (p. 59; cf. Gallagher 2003, p. 61). I want to focus my discussion of Gallagher's position on the following points.

First, I want to take a short look at what he means by 'body'.

The second and third point of my discussion pertain to the question of how Gallagher thinks that the body constitutes perceptual space. Two lines of argument in his work are supposed to shed light on the constitutive basis of spatial experience, focussing on proprioception and body schemata, respectively. However, I will argue that each strand, and also the relationship between both, remain dubious.

Physiology or proprioception?

Let me start with the first point regarding the meaning of 'body'. What does the above-cited claim that "the body provides the basis for an egocentric spatial frame of reference" mean? There are at least two possibilities, which Gallagher does not distinguish. This leads to problems in interpreting his account.[62] The two possibilities are: (1) the actual physiology of the body determines the egocentric structure of perception and (2) a representation of the body determines the egocentric structure of perception.

Passages like the following suggest alternative (1): "One of the important functions of the body in the context of perception and action is to provide the basis for an egocentric spatial frame of reference" (Gallagher 2005, p. 59). "The fact that perception is perspectivally spatial is a fact that depends precisely on an implicit reference to the spatiality of the perceiving body" (p. 137). Furthermore, the reference to Jakob von Uexküll (1920; Uexküll and Kriszat 1956) and to his conviction that perception is dependent on the physiology

62 This observation is shared by de Vignemont (2006), p. 7.

of the perceiving organism, in particular its sense organs and abilities of movement, suggests that Gallagher aims to explicate how the physiology of the body determines perception (Gallagher 2005, p. 140).

On the other hand, there are statements that suggest (2): "If the spatiality of action and the visual perception of the world are egocentrically organized by an implicit reference to our bodily framework, the *awareness* that is the basis for that implicit reference cannot itself be based on the egocentric framework of visual perception without the threat of infinite regress" (p. 60; my emphasis). "In its *prenoetic* roles the body functions to make perception possible and to constrain intentional consciousness in various ways" (pp. 138–9). These statements indicate that a representation of the body determines perception.

Both hypotheses are plausible, *prima facie*. However, they should be kept apart. Of course, physiology plays a role in determining the structure and possible contents of perception: humans can e.g. only see a distinct part of the colour and sound spectrum. The physiology of our bodies determines the kinds of perception we enjoy and the limits of what we can perceive within the modalities.

However, I think that Gallagher wants to defend (2). His central argument for the possibility of proprioception to ground the structure of exteroception (see below) would be superfluous if he thought that the mere physiology of the body was crucial.

I take it that Gallagher simply fails to delineate reading (1) from alternative (2). In what follows, I will assume that some representation of the body is supposed to constitute perceptual space. Let me now turn to the discussion of his argument to the effect that proprioceptive representations ground the egocentric structure of perceptual space.

Avoiding a possible regress

Gallagher wants to defend the thesis that a certain representation of the body constitutes the spatial structure of perceptual space. The relevant representation of the own body is proprioceptive representation. Since he takes proprioception as a kind of perception, Gallagher sees an immediate complication: If the representation of the body that is supposed to constitute the egocentric frame of reference in perception was perceptual itself, it must be ruled out that it is itself structured egocentrically. Otherwise, the thought that proprioception is the basis of egocentric exteroception would not be intelligible.[63] His conclusion is that „proprioceptive awareness [...] operates in a non-perspectival, intra-corporeal spatial framework that is quite different from the egocentric structure of action and exteroceptive perception" (p. 60).[64]

Let me turn to Gallagher's way of dealing with the threat of regress and of arriving at his conclusion.

Gallagher (2003) argues for a principal distinction between proprioception ("non-perceptual awareness") and exteroception ("perception"). He wants to defend the view that proprioception is not structured egocentrically, in contrast to exteroception. This argument is supposed to be necessary for the intelligibility of the claim that egocentric perception is grounded in proprioception.

Gallagher defines proprioception as "the bodily sense that allows us to know how our body and limbs are positioned" (2005, p. 43). He distinguishes two kinds of proprioception. On the one hand, proprioception can refer to the conscious shift of attention towards parts of one's own body – "reflective self-examination" (2003, p. 56). Since this kind of proprioception is similar to perception in that parts of the body are identified as objects and explicitly delineated from other things,[65] it cannot be the searched-for basis of other kinds of perception. Accordingly, Gallagher is after a different notion.

---

63 "If one accepts the premise that sense perception of the world is spatially organized by an implicit reference to our bodily framework the awareness that is the basis for that implicit reference cannot depend on perceptual awareness without the threat of infinite regress" (Gallagher 2005, p. 137). Gallagher accuses others (namely Antonio Damasio and Gibson) of not having seen this problem and of conceptualizing proprioception analogue to other forms of perception what would lead to the infinite regress and render their accounts unconvincing.

64 One might dispute that there is a thread of an infinite regress at all and claim that the egocentricity can be taken as primitive. Gallagher does not consider this possibility.

65 Whether this implies that this form of proprioception is egocentrically structured as one might suppose and which consequence this would have for his account in which 'egocentric' is a synonym for 'body-centred' is left open here.

This other form of awareness of the body is supposed to be an unconscious, implicit grasp of the pose and position of limbs that enables the control of actions (p. 59). The important point he needs to show is that the spatial relationships are not represented egocentrically in this case. According to Gallagher, in proprioception there is no perspective on the body, no point of reference, relative to which spatial relations are specified (Gallagher 2003; and 2005, pp. 137–8).[66] Whereas in exteroception, distances and directions are represented relative to the body, in proprioception, spatial relationships are represented *absolutely*. This is supposed to be the reason why it does not make sense to say that one part of the body is further away than another one. Nor is the right hand ever regarded as an additional left hand if it is held at the corresponding position. He sums up his view pointedly:

> Generally speaking, the proprioceptive spatiality of the body is not framed by anything other than the body itself. In other words, proprioception is a non-perspectival awareness of the body. (Gallagher 2003, p. 62)

What he means to argue, that proprioception is "framed by the body", can be made clear by distinguishing between two ways in which bodily locations can be specified, namely A-locations and B-locations.

A-locations and B-locations

Bermúdez distinguishes two different ways in which one can specify a given location on the body (Bermúdez 1998, p. 153–61; 2003; and 2011, p. 174–7). The following example from Bermúdez helps to reveal the contrast:

> (1) I have a pain at a point on the sole of my right foot when I am standing up, and my right foot is resting on the ground in front of me.
>
> (2) I have a pain at the same point on the sole of my right foot when I am sitting down, and my right ankle is resting on my left knee. (Bermúdez 1998, p. 154)

66 He refers to O'Shaughnessy (2008) and Bermúdez (1995) and regards himself being in the tradition of Merleau-Ponty and Jean-Paul Sartre.

Do I feel the pain at the same location in both scenarios? Bermúdez argues that this question has no definite answer since one has to distinguish two kinds of location, A-location and B-location. The answer to the question depends on the kind of location one takes into account.

The difference of the two kinds of location is based in the fact that the human body is made up of moveable parts that are put together via joints. A-location is specified by its distance and direction to the respective nearest joints. B-location is specified with reference to the joints that lie between the torso and the limb on which the point in question is located.

Applied to the example, the A-location is identical in (1) and (2), whereas the B-location is different in both scenarios: the location of the pain is the same in both cases relative to my right ankle; it is different, however, relative to the torso in both cases.

Gallagher can be interpreted as referring to A-locations in claiming that proprioception employs an absolute frame of reference. This means that locations represented in this way stay the same despite bodily movements. Therefore, he claims, "proprioception follows the contours of my body, but not from a perspective" (Gallagher 2003, p. 63).

Some critical remarks regarding the avoidance of the infinite regress

Let me note some critical comments regarding the way Gallagher wants to avoid the infinite regress.

First, I it should be noted that the distinction of A- and B-location is not distinctive of the human body. It can be applied to many other objects, too. In a certain sense, the rust on the door of my car changes its location when I open the door, namely with respect to the remaining parts of the car (analogue to the B-location); but it remains at the same location in another sense, namely with respect to the hinges of the door (analogue to the A-location). In allowing for the distinction between A- and B-location, the representation of the body is not fundamentally different from representations of other objects that are made up of moveable parts.

Furthermore, the distinction between A- and B-locations seems to be a mere difference in frame of reference. However, it is nothing extraordinary that the same moving point changes its location or not depending on the frame of reference relative to which it is specified. This applies to any case of moving locations: for example, the city of Munich does not change its location relative to other cities in Germany although it changes its location relative to cities in North America due to the movement of the Eurasian plate.

These two comments emphasize that the proposed spatiality of proprioception is no spatiality *sui generis* of the representation of human bodies. The analysis of different ways in which locations might be represented is applicable to other objects beyond human bodies. Hence, we should not expect too much of an explanation for the peculiar nature of body representation from stating these features.

Third, the foundation by (absolute) A-locations seems to be problematic in that thereby no integrated model of the body can be established. If bodily locations are specified with respect to the nearest joints, all that is established is a bunch of unrelated bodily locations. However, their respective relation remains unclear. This latter information is only given by B-locations. However, these are not available to Gallagher's conception of proprioception since they specify locations to a certain point of reference, namely the torso, and thereby count on his view *per definition* as egocentric; but to avoid the infinite regress he needs absolute positions.

Fourth, in Gallagher's account bodily space and environmental space are separated. One might wonder whether the assumption of an integrated frame of reference in which both bodily positions and positions of objects in the environment are represented is more plausible for e.g. the explanation of goal-directed acts. On this proposal, computations of converting different kinds of locations that might complicate and slow down the control of movements would not be necessary. On Gallagher's view, however, something like this seems necessary.

Fifth, it is not entirely clear how talk of 'left' and 'right' is possible without assuming any points or axis of reference. Gallagher himself, in explicating proprioceptive awareness, talks about the "left [!] hand" not becoming a right one, if I hold it near the right shoulder (2003, p. 63). However, it is unclear how the talk of left and right with regard to absolute positions can be understood.

These points may not undermine Gallagher's position entirely. However, they show that there are some caveats to his account. I want to leave these issues aside though and discuss the way that proprioceptive awareness is supposed to ground egocentric perceptual space on Gallagher's view.

Constitution of egocentric space

Unfortunately, the search for an explicit answer to the question of how bodily awareness constitutes egocentric space in Gallagher's work remains unsuccessful. While he develops the first part of the argument (how to avoid the infinite regress) in detail, the remarks on the second part about the constitutive relation between bodily awareness and perceptual space are quite vague (Gallagher 1995; 2003; 2005; 2009).

Gallagher (2009) at least shows how the constitution is *not* to be conceived. He criticises a study of Asifa Majid et al. (2004), who propose that the egocentric frame of reference is to be regarded as derived from an allocentric, object-centred frame of reference. Their basic idea is that one can derive the egocentric frame of reference from parsing the subject's body according to its internal spatial and functional organisation. The parsing results in a set of axes that are projected beyond the body and make up the frame of reference for perception. This establishes a frame of reference that allows judging other things as being located left, right or in front of the subject (p. 109).[67]

Gallagher rejects such a derivation of the egocentric frame of reference from an allocentric, object-centred frame. On his view, the egocentric frame of reference is the basis of all other frames of reference (Gallagher 2009, p. 354) and not itself based on any other frame of reference.

67 Their position is strongly reminiscent of the view that is criticized in Campbell (1994), pp. 9–10.

Practical engagement and body schemata

How does Gallagher himself think about the constitution of the egocentric frame of reference? Gallagher (2005), esp. pp. 138–52 gives some hints. Interestingly, the spatiality of the subjects' body plays only a secondary role here. Instead, he focusses on the subject's practical engagement with the world. Gallagher refers to Uexküll, Merleau-Ponty, Gibson and others. The key element is the body as represented in the body schema.

> The body projects a pragmatic and egocentric spatial framework that does not correspond precisely to objective (allocentric) spatial measurements. Body schemas enable us to find our way in space; to walk without bumping into things; to run without tripping and falling; to locate targets; to perceive depth, distance, and direction; to throw and to catch a ball with accuracy. (p. 141)

The egocentric spatial frame of reference is obviously tightly related to the body schema that, as part of the prenoetic bodily awareness, is constitutive of the spatial structure of perception.

What does Gallagher mean by 'body schema'? Gallagher describes the body schema or schemas[68] as those unconscious sensorimotor processes that enable the subject to interact dynamically and flexibly with the environment based on its intentions. "The body schema [...] involves certain motor capacities, abilities, and habits that both enable and constrain movement and the maintenance of posture" (p. 24; cf. pp. 32–4; pp. 37–8). "[T]he body schema reflects a practical attunement of the body to its environment" (Gallagher 1995, p. 239; cf. pp. 226–239).

These quotes make clear that the body schema comprises a range of basic capacities of how the embodied subject can react toward stimuli or accomplish basic movements, thereby enabling the organism to interact with its environment. One important property of the functioning of the body schema for Gallagher is that it operates, in principle, on a pre-conscious level. This is the key feature that distinguishes it from the body image that he takes to be

68 The plural 'body schemas' refers to particular motor programmes that each are defined by a particular pose or movement like the movement of hand to mouth; cf. Gallagher (2005), p. 24, n. 5. Gallagher uses both, the singular and plural version.

available to consciousness (Gallagher 2005, p. 26).[69] The body schema is thus a "system of sensorimotor functions" that involves "tacit performances – preconscious, subpersonal processes that play a dynamic role in governing posture and movement" that works "close to automatic" (ibid.).

Gallagher's claim is that this body schema is the central feature that "shapes" consciousness and, in particular, perception.

> In its *prenoetic* roles the body functions to make perception possible and to constrain intentional consciousness in various ways. Here are included operations of body schemas – the body's non-conscious, sub-intentional appropriation of postures and movements, its incorporation of various significant parts of the environment into its own organization. (p. 139)

One difficulty with Gallagher's argument here is that in describing the role of the body schema with regard to perception, he does not sharply distinguish between the claim that the working of body schema has an effect on conscious perception and the claim that it is constitutive of conscious perception. In general, he seems aware of the difference: on another occasion, in stating the impact that emotions have on perception, he distinguishes between both understandings (p. 151). The reading there suggests that he uses 'shapes' for 'being constitutive of'. Nevertheless, in describing how the body schema relates to experience, he frequently speaks about the "effects" that the workings of the body schema have, or how the body schema "enables" the subject to perceive.

With regard to the important distinction between instrumental and constitutive dependence, Gallagher's writings are thus ambiguous. Speaking of effects that the body schema has on perception might indicate a rather instrumental relation to perception instead of a constitutive one. However, I take it that he wants to defend a constitutive claim, but that he fails to delineate both dependencies sharply. This results in mixing up examples for instrumental and constitutive dependence of perception on action (pp. 139–40; pp. 144–5).

69 However, the functioning of the body schema might be altered by the body image or other conscious states (Gallagher 2005, p. 27). See ibid., pp. 24–30 for more details about how he distinguishes between body image and body schema.

Summary of Gallagher's account

Let me look on Gallagher's overall account. He develops two lines of thought to explain spatial perception. On the one hand, Gallagher puts a lot of effort in arguing for the distinctive non-perspectival way the body has to be represented in proprioception for it to be the basis of exteroception. On the other hand, he states that the basic practical bodily capabilities that make up the body schema are what constitute egocentric perception. Gallagher does not link the two claims, however. Moreover, it is dubious if it is possible to do so.

In particular, from Gallagher's own theoretical perspective, one would expect that the body schema be linked to proprioception in important respects. Why would he have to argue for the claim that proprioception does not employ an egocentric frame of reference at all if proprioception were not relevant for the functioning of the body schema? However, surprisingly, regarding the relationship between body schema and proprioception, he says nothing beyond the claim that proprioceptive information is part of the input that is used by the processes of the body schema (pp. 37–8).

His basic conviction seems to be that a prenoetic bodily awareness constitutes the egocentric structure of perception. The body schema is a central part of this bodily awareness. The latter comprises basic sensorimotor programmes and abilities that enable basic goal-directed coordinated movements. These include spatial information about the position of limbs given in proprioception.

If this is right, a central statement of Gallagher's has to be revised: "The fact that perception is perspectivally spatial is a fact that depends precisely on an implicit reference to the spatiality of the perceiving body" (p. 137). In order to get the actual account of Gallagher, it is necessary to add the body schema, which is a system of capacities on his understanding.

What has now become clear is that, on Gallagher's account, the egocentric spatial structure of perceptual experience is not just based in some representation of the spatial structure of the body in proprioception as he claims here and there. In contrast, the body schema, understood as a collection of basic motor capabilities that draws on proprioceptive information, is the essential element in constituting egocentric space.

However, the relationship between the non-perspectival proprioception and the body schema is dubious in Gallagher's framework. He does not elaborate on the exact relation between the two. In addition, there seem to be some issues in spelling out this relationship given his views. One might suppose that a body representation in absolute co-ordinates is not useful for the functioning of the body schema. Such a representation neither includes the relative positions of different body parts nor is it clear, as stated above, that it is *one* body that is so represented. This seems to be in contradiction with the supposedly holistic character of the body schema and thereby with a basic condition on the possibility that the body schema can fulfil its supposed function. Take for example the maintenance of balance that Gallagher uses frequently as an illustration for the working of the body schema. To be able to maintain balance depends on the simultaneous coordination of very many body parts. Changes in some parts have to be compensated for by changes in other parts. Without this holistic organization of the body schema, a high-wire act or even upright gait would be impossible. The holism is an integral part of the body schema. However, it is difficult to explain it with the notion of proprioception that Gallagher proposes. Regarding these questions there are large caveats in his conception that might be avoided somehow but, as far as I can see, his work does not offer any solution.

In sum, Gallagher's position only pretends to ground the nature of spatial perception in the represented spatiality of the body alone. It is doubtful after this discussion whether this would be possible at all. Ultimately, the explanatory weight is carried by the practical engagement of the embodied subject with its environment, subserved by the body schema. However, Gallagher is silent about the details of the constitutive relationship between practical 'attunement' to the environment and perceptual space. Nevertheless, this idea makes his position akin to positions like that of Merleau-Ponty, to which I now turn.

2.2.4.2 Embodied action II – Maurice Merleau-Ponty

The basis of my discussion is Maurice Merleau-Ponty's *Phenomenology of Perception* (2002).[70] In contrast to Gallagher, Merleau-Ponty spells out in more detail how the body is supposed to constitute the spatiality of perception. I will mainly focus on two sections of Merleau-Ponty's monograph: "The spatiality of one's body and motility" (Part I, Ch. 3) and "Space" (Part II, Ch. 2).

Overview of the essential issues

I take the following three issues as most important with respect to Merleau-Ponty's account of the constitution of spatial perception. The explication of them will result in a firm understanding of the basic tenets of his view.

First, since the body is all-important, I will clarify the understanding of the body within the account. Merleau-Ponty, inspired by Husserl, distinguishes two meanings of 'body'. On the one hand, the body figures as a physical object in the world that is similar to all other material objects ('body'). On the other hand, the body figures as that by which one perceives, feels, acts, in short: lives ('living body').

Second, I will explicate how, according to Merleau-Ponty, the living body is supposed to constitute perceptual space. Central to this point will be the notions of figure and ground, 'gearing' ('la prise'/'Anhalt') and, again, the body schema.

Third, I will also look at some of the empirical evidence Merleau-Ponty puts forward in support of this view.

The living body

Let me start with the notion of the living body that is key to Merleau-Ponty's account. He claims, "far from my body's being for me no more than a fragment of space, there would be no space at all for me if I had no body" (Merleau-Ponty 2002, p. 117). In which sense does he refer to the body in this quote?

70 This edition was first published in 1962. (However, page numbers are not identical in both editions.) I will also rely on Merleau-Ponty (1945), and (1974).

As indicated, Merleau-Ponty sharply distinguishes between two different notions of 'body'. In this quote, he does not refer to the body in the sense of a mere object, located in the world as other objects. This way of using 'body' is appropriate in some cases. In a certain sense, human bodies are just objects situated in space and time just as any other object. However, this is not the sense of body that is meant here. The body that grounds perception is the living body, the body regarded as that by which we perceive and act. In addition, this one can never be an object of perception. Hence, it is not the body as the thing one can see in a mirror, or touch. It is not the body as a possible object of perception, investigation or action. It is the body as it figures in the acts of perceiving, investigating and acting. It is the body in its function to enable one to get in contact with the world. It is the body as enabling me to live in the world.[71] The following quote summarizes the distinction nicely:

> [T]he first is a system of bones, muscles and flesh brought down at a point of space, the second shoots through space like a rocket to reveal the external object in its place. In so far as it sees or touches the world, my body can therefore be neither seen nor touched. What prevents its ever being an object, ever being 'completely constituted' is that it is that by which there are objects. It is neither tangible nor visible in so far as it is that which sees and touches. The body therefore is not one more among external objects [...]. (p. 105; cf. p. 121, p. 177)

71 Merleau-Ponty uses the French word 'corps' throughout. In German, it is possible to express the difference with the distinction between 'Körper' and 'Leib'. No such linguistic contrast is available in French or English. I will use the term 'living body' for 'Leib', where it is not evident from the context. Otherwise, I will simply use 'body'. Merleau-Ponty's understanding of the body is based on, but not the same as Husserl's. Cf. e.g. Husserl (1991), pp. 161–2, where he states that the body is distinguished essentially from all other things: "Einerseits ist der Leib auch ein Ding, physisches Ding wie irgendeines sonst [...]. Es ist ein Ding unter anderen Dingen, zwischen ihnen hat es eine wechselnde Lage, es ruht oder bewegt sich wie andere Dinge. Andererseits ist dieses Ding eben Leib, Träger des Ich; das Ich hat Empfindungen, und diese Empfindungen werden im Leib ‚lokalisiert' [...]." ("On the one hand, the body [Leib] is a thing, too, a physical thing as any other. It is a thing amongst other things, it has a changing location among them, it stays or moves as other things do. On the other hand, this thing *is* body [Leib], bearer of the I; the I has sensations, and these sensations are localized on the body [...]"; my translation.) For a comparison between Husserl's and Merleau-Ponty's concept of the body cf. Carman (1999).

One might ask whether the distinction that is made here is an ontological one or merely one of perspective (cf. Baldwin and Bell 1988). Jean Paul Sartre made a similar distinction – between the 'body-for-oneself' and 'the body-for-others' – meant as an ontological distinction where both notions refer to different objects. However, this does not seem to be Merleau-Ponty's view. It is rather a distinction between two different ways of presentation of the same object. In a footnote he speaks about "two views of me and my body: my body for me and my body for others" (p. 131, fn. 17). This clearly is reminiscent of Sartre's distinction without following the ontological implications. Hence, what is crucial in his account is the body for me, the living body, as it functions to perceive and act on the world. However, this does not mean that this body is ontologically separate from the body as an object in the world. It is this same body but conceived in its largely prereflective and preconscious functioning that constitutes our way of perceiving and acting. How does Merleau-Ponty cash out this idea?

The body schema

The most important notion in this context is the notion of the 'body schema'.[72] This becomes clear at the beginning of the chapter on "The spatiality of one's body and motility" where Merleau-Ponty contrasts the body and the living body, the latter of which is present in the body schema. The notion of body schema already appeared in the discussion of Gallagher's account. In Gallagher, the body schema referred to basic preconscious capabilities of bodily movements, such as motor programmes that enable fluent and flexible behaviour. The usage in Merleau-Ponty is slightly different, however.

72 Let me add one important note right at the beginning. As already noted in the discussion of the account of Gallagher there are two kinds of body-related notions that are important and related but have to be sharply distinguished. These are 'body image' and 'body schema'. Merleau-Ponty speaks of "schéma corporel" (e.g. Merleau-Ponty 1945, pp. 114–120). Thus, he intends the latter. He refers to Conrad (1933). Unfortunately, Smith's edition of Merleau-Ponty's *Phénoménologie de la perception* does not reflect this difference. He translates "schéma corporel" as "body image". However, one must admit that the distinction between the two notions then was not the same as it is today. Thus, it is especially important to have a close look at Merleau-Ponty's understanding of the notion.

Merleau-Ponty takes the body schema neither as a set of associations of different representations of the body, a "name for a great many associations of images" (p. 113; cf. Head and Holmes 1911; Pick 1922; Schilder 1923). Nor does he understand it as a "total awareness of my posture in the intersensory world" (p. 114).

On his view, quite similar to Conrad's (1933), the body schema is a special kind of Gestalt. Being a Gestalt essentially means that it is not an aggregate of particular bits of representations or information but a whole that cannot be reduced to the sum of its parts. Thus, the whole is more basic than the parts (both epistemologically and ontologically). The body schema is a *special* Gestalt in that it is fundamentally determined by actual or possible tasks, it is "dynamic" (Merleau-Ponty 2002, p. 114). Through the body schema the living body is always already oriented towards (possible) goals, is always already engaging with the world. Thus, as Merleau-Ponty puts it, "the body schema is finally just a different word for saying that my body is toward-the-world"[73]. Furthermore, I know of my limbs and their position through the body schema that encompasses them all (pp. 122–3).[74]

In effect, the body schema has three functions. First, it grounds a special kind of awareness, in which my body is presented to me as a Gestalt. Second, it is the basis of my knowledge of how my body is situated in space, how my limbs are spatially arranged. Third, it is a condition on my ability to interact, to engage with the world, which is expressed in the claim that the body

73 This is my own translation of Merleau-Ponty (1974), p. 126 (Merleau-Ponty 1945, p. 117). Smith translates it as "the body image is finally a way of stating that my body is in-the-world." (Merleau-Ponty 2002, p. 115). I already mentioned the difference between 'body image' and 'body schema'. Being-in-the-world is a term that is introduced by Heidegger (2010). Merleau-Ponty's "being-toward-the-world" might be similar but it is in an open question whether one can simply identify both notions. I do not want to discuss this issue here.

74 This is a difference to Gallagher's account, where knowledge of the positions of the limbs is based on the body image.

schema is just another expression for "*être au monde*" (being-toward-the-world).[75]

However, is there more to be said about the body schema than just stating its functions? What *is* the body schema essentially? To find an answer to that question, it is useful to have a short look at Kant, where schemata play an important role in the explanation of perception (cf. Carman 1999, pp. 218–9). Kant 1998, pp. A137–47/B176–87 tries to answer the difficult question[76] of how it is possible to apply "pure concepts of the understanding to appearances in general" (p. A138/B177). Recall that for Kant, "[t]houghts without intuitions are empty, intuitions without concepts are blind" (p. A51/B75). Hence, the interplay of concepts and appearances is one of the core ideas in Kant's philosophy.[77] Therefore, he needs an account of how both connect, since they belong to fundamentally different realms. Here is the point where the schemata come into play as a "general procedure of the imagination for providing a concept with its image" (pp. A140/B179–80), as the "pure synthesis, in accord with a rule of unity according to concepts in general" (A142/B181). Thus, the schema essentially is a rule or procedure that determines the application of categories to particular appearances or the particular imagination of general categories that is done by the power of judgement.

While Kant thought that the schemata are generated by the faculty of imagination, Merleau-Ponty sees the body schema realised in the habitual engagement of the living body with its environment. Like Kant's schemata, Merleau-Ponty's body schema has its place between sensation and mature perception, grounding the applicability of one to the other. The body schema, which is nothing beyond the ability of the habitual but flexible, non-reflective and preconscious engagement with the world that is given with the body, has

75 Tiemersma (1982), p. 253 summarizes these points in stating that the body schema "is a very rich one, incorporating aspects of practical intention, expression and comprehension".

76 "This schematism of our understanding with regard to appearances and their mere form is a hidden art in the depths of the human soul, whose true operations we can divine from nature and lay unveiled before our eyes only with difficulty" (pp. A141/B180–1).

77 That is why it is no surprise that Kant took the chapter on schematism to be one of the most important of the "Critique of pure reason"; cf. Kant (1998), Editorial notes, p. 728, n. 51.

a similar theoretical position as the schemata in Kant's philosophy have. They are not only similar in their fundamentality in both approaches, but also in their primitiveness: both philosophers argue that more cannot be said about them than stating their functions.

Interim conclusion

Let me take stock and look at what I have said so far about Merleau-Ponty's account. I started with a quote that makes clear that Merleau-Ponty claims that without a body there would be no perceptual space. I then made clear what is meant with the 'body' in this context, namely the 'living body'. I outlined what is distinctive about the living body, which is essentially captured in the various functions of the body schema. Ultimately, and this is an extraordinary claim of Merleau-Ponty's, the body is taken by him as a subject in its own right: "[T]he body is a natural self and, as it were, the subject of perception" (Merleau-Ponty 2002, p. 239). This claim is essentially directed against views such as Kant's which he took to be overly intellectualistic. There is no need for a transcendental ego; the living body does the trick.

These are the cornerstones of the notion of the body as it figures in Merleau-Ponty's account. In what follows, I will turn to the question of how the body constitutes perceptual space.

'Gearing'

Merleau-Ponty's account of how perceptual space is grounded centres on the idea that the living body is always *geared* to the world.[78] This idea, which is central but difficult to understand, might be spelled out as follows.

I think one can conceive of gearing as another end of explanation, another basic fact that comes with the body schema and cannot be spelled out in any more primitive terms but can only be described. The gearing to the world is an ever-present constituent of human embodied nature. Body schema and

78 In the original, he uses "la prise". The German edition has "Anhalt". One might trace back the notion of gearing to the writings of Heidegger. As in Merleau-Ponty, the notion of gearing is strongly connected with the idea of human being-in-the-world as an *oriented* mode of existence; cf., e.g., Heidegger (1979), p. 319.

gearing to the world are aspects of the being-toward-the-world that human existence essentially consists in. The fact that we are always and constitutively geared to the world is a shorthand for saying that we are always oriented toward the world according to our intentions and bodily needs; that the world, on the other hand, is perceived against the background of practical engagement and structured by it. Indeed, Merleau-Ponty claims that "[w]e remain physically upright not through the mechanism of the skeleton or even through the nervous regulation of muscular tone, but because we are caught up [*nous sommes engagés*] in a world" (Merleau-Ponty 2002, p. 296, n. 18; cf. Merleau-Ponty 1945, p. 294, n. 19).

The idea of gearing essentially means that the world is a kind of handhold for us. This is terminologically present in Merleau-Ponty's usage of 'la prise', which literally means 'grasp'. It also comes out in the German 'Anhalt' that is associated with the idea of something that helps to maintain balance. The English 'gearing' is slightly different in meaning, although not inappropriate: the world and the living body engage with each other as cogwheels in a gearbox do; they are highly attuned to each other. This is different from the idea that the world helps to maintain our balance. However, it is in accordance with the body schema and its functions that enable our fluent and flexible engagement with the world, which usually presents itself to us as a space of easy-to-accomplish possibilities ready for us to realise. In that sense, we are attuned with the world, which is well brought out by the term 'gearing'.[79]

With these rather general considerations about how Merleau-Ponty understands perception, I now turn to the special case of perceptual space and its explanation. It will come out that all elements mentioned before are part of the explanation.

The constitution of perceptual space

The crucial passage for the explanation of the constitution of perceptual space is the beginning of the second chapter of the second part, which is headlined

79 Merleau-Ponty describes the relationship between the bodily activity and perception in terms of figure and ground, too. Originally, the structure of figure and ground means that everything one perceives is always perceived as a figure against a certain background, an idea that was developed by Gestalt psychologists; cf. Merleau-Ponty (2002), p. 4; p. 117.

"Space". In that section, Merleau-Ponty draws on experiments employing various methods to examine the long-term effects of perceptual distortions. Merleau-Ponty takes these studies to be indispensable for examining the constitution of perceptual space (Merleau-Ponty 2002, p. 284).

He (pp. 285–293) cites experimental results from Stratton (1897), and Wertheimer (1912). Stratton used inverting goggles that make the world appear upside down. According to his report, the initially upside-down turned visual field rights itself after some days. On Merleau-Ponty's interpretation of the findings, this process comprises effectively three stages. First, the visual image is turned upside down, everything seems unreal. Second, from the second day onwards the world is perceived upright again but now the own body seems to be either turned or in an abnormal position. In the third stage, the body increasingly rights itself, particularly when the subject is active. Notably, successful goal-directed movement is not possible before the fifth day and unseen parts of the body remain experienced in their former spatial position. The correct localization of sounds is not possible until the eighth day and only when the source of the sound is simultaneously seen.

The second study Merleau-Ponty discusses is reported in Wertheimer (1912). In this study, the perception of a room is distorted by a mirror so that the room appears to be tilted 45 degrees off the vertical. When subjects are exposed to this situation (they can only see the mirror image), they see the room initially tilted. However, there is a sudden re-instantiation of a 'normal' perception after a few minutes: the room appears straight up again. In this case, the change happens without any preceding active exploration or interaction of the subject with the perceived environment as was necessary in the inverted vision case.

How does Merleau-Ponty interpret these experimental results? In both experiments, there is a conflict elicited between differently oriented sensational fields. In the first case, the conflict holds between synchronously present sensational fields – visual and tactile. In the second case, the conflict is diachronically between two visual fields – the one that was present before subjects looked in the mirror and the one that is present while they are watching the scene in the mirror. Both conflicts vanish, though after different amounts of time. The crucial question is how the sudden or increasing change of orientation of the visual field is to be explained.

For understanding Merleau-Ponty's answer to this question, the notions of living body, body schema, and gearing that I discussed above are essential. For him, the notion of gearing is crucial: what marks the transition from a perception of a scene as tilted, or disoriented to a perception of an oriented, 'normal' world is the transition from initially not being geared to the new environment to being geared to it. Abnormal perception is based on one's being geared to the world one used to perceive before putting on inverting goggles, or looking in the mirror. That is why the new perception seems strange, and impossible to rely on in acting. However, time or/and habituation lead to a change in that one gets geared to the newly presented environment which in turn 'rights up' perception.

Thus, our 'normal' upright perception is nothing beyond a perceptual correlate of the gearing of the body to the world.

Spatial perception of the environment depends on the way the body is geared to the world. I already explicated that the gearing consists essentially in the way body and world are linked by the body schema. Thus, the way the world appears to us spatially is constituted by the body and its habitual engagement with the environment that is generated beyond our conscious control. Merleau-Ponty articulates this thought in the following way.

> There is [...] another subject beneath me, for whom a world exists before I am here, and who marks out my place in it. This captive or natural spirit is my [living] body, not that momentary body which is the instrument of my personal choices and which fastens upon this or that world, but the system of anonymous 'functions' which draw every particular focus into a general project. [...] Space and perception generally represent, at the core of the subject, the fact of his birth, the perpetual contribution of his bodily being, a communication with the world more ancient than thought. (p. 296)

2.2.4.3 Summing up embodiment approaches

Both accounts that I have discussed in this chapter take the body as a constitutive condition of perceptual space. Unfortunately, Gallagher's account lacks clarity on most important issues. Three lines of thought are mixed up in his account somehow: first, an argument that centres on the physiology of animals in a von Uexküllian fashion; second, an argument to the effect that non-perspectival proprioception constitutes perceptual space; and third, another argument to the effect that 'prereflective' functions of the body schema are what explains perceptual space. I tried to make sense of his account in proposing a possible understanding of how these strands of thought can be integrated. However, the argument focussing on the non-perspectival proprioception remains especially unclear (not only with regard to how the argument is supposed to work, but also with regard to its role in the overall argumentation, given the later emphasis on the body schema). In the end, one has to concede that Gallagher's position has many open ends and it is difficult to see what his arguments really are. In Gallagher's work, one can find a lot of notions and ideas that are present in Merleau-Ponty's *Phenomenology of Perception.* I suspect that he wants to defend a similar account of perceptual space as the French phenomenologist.

However, Merleau-Ponty has a clearer way of arguing. The central concepts in his approach are spelled out more carefully. The practical engagement with the world that is marked by the concept of gearing is the central explaining element when it comes to perceptual space. For Merleau-Ponty, this gearing is tied to the living body. He takes the living body as an autonomous subject that has ways of receiving information and acting that are beyond the person's conscious dealings with the world. Only examining strange situations or pathologies can reveal these fundamental interactive structures on which our personal conscious perception and experience depends. Thus, for Merleau-Ponty, perceptual consciousness is built on the fundamental bodily processes that might be seen as residuals of our phylogenetic development that were already there when consciousness and a 'personal life' was not. Surely, in the realm of the person with its consciousness, self-consciousness, rationality and thinking we are able to go beyond these ancient and immediate ways of coping with the world. However, Merleau-Ponty claims that all these sophisticated ways of perceiving, thinking about and acting on the world are fundamentally and inextricably linked and built on a subject that belongs to us but is largely autonomous: our living body. In particular, one of the crucial conditions for all intellectual operations we are able to perform to pursue our goals in this world, that is perception, is claimed to be constitutively dependent on how our living bodies are situated in and interacting with the world. Thus, Merleau-Ponty's is an account in which the body in its pre-personal interaction with the world constitutes the nature and structure of perceptual space.

2.3 Systematic overview of constitutive accounts II: *Kinaesthesis*, potentiality and disposition to act

In this chapter, I want to consider three positions in which perceptual space is thought to be constitutively dependent upon the dispositions, potentiality or capability to behaviour that subjects have. In contrast to Noë's account, it is not some knowledge about sensory consequences of (self-)movement that grounds perceptual space. Instead, similar to the embodiment approaches I discussed before, it is claimed that perceptual space constitutively depends on the practical possibilities the perceiving subjects have.

The idea is cashed out in different ways by these thinkers. Husserl's central thought is that the whole system of own practical possibilities constitutes spatial perception. In Scheler's account, the grasped potentiality to self-induced movement constitutes perceptual space. Evans claims that dispositions to behaviour partly constitute the egocentric spatial contents of perception.

However, there is also an important difference to the embodiment approaches. Whereas in the former accounts claims about the living body and its engagement with the world figure centrally in the explanation of perceptual space, the positions of the late Husserl, Scheler, and Evans are not dependent on claims about embodiment. On these accounts, only dispositions to behaviour matter for an explanation. These authors seem to think that the fact that the practical dispositions are realized as bodily processes does not add something substantial to the explanation of the nature of perceptual space. They might be realized otherwise.

My discussion of these approaches will be structured as follows. The accounts are introduced with respect to the core claims about how dispositions to action are taken as a constitutive condition on perceptual space. A second aim is to show their independence from claims about embodiment. None of the three positions treats this question explicitly. Nevertheless, any of these accounts contains some thoughts that make it plausible that what essentially explains perceptual space is something that is not tied to an embodied nature of the subject. Therefore, the body seems dispensable in the explanation of perceptual space.[80]

2.3.1 Late Edmund Husserl

Husserl's account is centred on the concept of *kinaesthesia*. This is true for every stage of his philosophical thinking. However, as already indicated, only the term stays the same while its meaning changes significantly. I already presented some steps of the development of this notion in Husserl's thinking (see sections 2.1.1 and 2.2.2 above). Here, I will now only explicate his understanding in the last period of his philosophy.

In his later work, Husserl distinguishes between two different functions or roles that *kinaestheses* have. On the one hand, the term refers to the actual movements we perform. Viewed in this regard, they are termed "practical functioning" ("praktisch fungierend"). On the other hand, they are "perceptual functioning" ("wahrnehmend fungierend") in which role they are constitutive of perceptual space (Husserl 2008a, pp. 396–7). He does not claim that

80 I discuss the implications on embodiment mainly to mark the substantial difference to the accounts discussed in the previous chapter. I do not want to claim that we can or should abandon any claims about embodied processes in a theory of perceptual space. Nevertheless, it is a mark of these theories that they try to formulate their views in ways that abstract from bodily processes.

the set of *kinaestheses* is divided into two different kinds but only that they serve two fundamental different functions.

All *kinaestheses* together form a system of possibilities ("Vermöglichkeiten"; Husserl 1954, p. 164, cited in Ferencz-Flatz 2014, p. 38).[81] Thus, *kinaesthesia* does not only mean sensation of movement or actual movement anymore but it refers to the whole set of possible movements. Husserl claims that this diachronic manifold of practical possibilities constitutes perceptual space (Neumann 1999, pp. 159–60; Claesges 1964, p. 73, p. 82).

Although it is true that Husserl conceives of *kinaesthesia* primarily as the system of possible *bodily* movements, one might wonder whether the body is not dispensable, since the kinaesthetic system is just a set of *possibilities*. Furthermore, in his later thinking, Husserl takes the kinaesthetic system not as the set of possible *overt bodily* movements. Instead, the kinaesthetic system consists of movements that belong to the inner perspective one has of the own body (Ferencz-Flatz 2014, p. 39; Husserl 1954, p. 164). He speaks of "inner mobility" ("innere Beweglichkeit") in contrast to the outer observable physical movements (Ferencz-Flatz 2014, p. 39). Although the inner movements correlate with observable bodily movements they are not reducible to them. Furthermore, they are also not just complexes of sensations. It is not entirely clear what these inner movements really are. It seems that he refers to the movements as experienced from the inside.

Given these claims, it is no big step to dispense with claims about the body altogether. Indeed, Husserl considers this possibility in stating that he sees no reason why being an embodied organism should be necessary for the existence of *kinaesthesia*.

> Aber denkbar ist es, dass ich gar keinen Leib habe, dass eine Natur ohne einen Leib sich durch Kinästhesen, Nah- und Fernkinästhesen, konstituiert. Ich kann

81 Cf. the statement from Husserl quoted in Ferencz-Flatz (2014), p. 36: "*Kinaesthesis* differs from *Empfindung* (sensation) by having an intimate relation to subjective potentiality" (originally in Cairns 1976, p. 4).

> nicht einsehen, warum das nicht möglich sein soll. [...] Es könnte sich doch rein kinästhetisch das Null konstituieren.[82] (Husserl 1973, p. 547)

By taking *kinaestheses* as bare possibilities, it is now possible for him to consider that it might be possible that the nil, which is the centre of perceptual space, might be constituted by *kinaesthesia* alone, without presuming a body; that is, space literally becomes a synonym for the set of possibilities one has.[83]

I want to close my discussion of Husserl's later account with these considerations. To get a comprehensive view of his thought is notoriously quite difficult since his thoughts are contained in thousands of pages of manuscripts, a lot of which are not published yet. However, I think it is justified to say that Husserl's account of the constitution of perceptual space draws on the practical possibilities we experience from the inside and that it allows for the claim that a constitutive explanation of perceptual space does not need to rely on claims about the embodied nature of the perceiving subject.

2.3.2 Max Scheler

Max Scheler's account is primarily articulated in his essay "Idealismus – Realismus" (1928)[84]. Here is the baseline of his explanation of space in few sentences, to get an overview of what will be discussed in more detail below. On Scheler's account, spatiality is constituted solely by the grasped power to self-induced movement. A particular space is a special kind of spatiality that is tied

---

82 „However, it is conceivable that I do not have a living body, that a nature can be constituted by *kinaestheses*, near and distal *kinaestheses*. I cannot see why this should not be possible. [...] The nil might be constituted purely *kinaesthetic*" (my translation).

83 Peacocke (2014) has also recently developed the idea that the basic entities that can be regarded as subjects, i.e. those that employ his Degree 0 and 1 of self-representation, can be explicated without taking into account notions of embodiment. He describes the experience of a subject with Degree 0 of self-representation as having a centre, a here, without thereby being a form of self-representation since there is no element in the content that refers to the subject itself. The 'here' is just a location (p. 31). This shows at least that the question whether subjectivity is essentially embodied is still debated in contemporary philosophy.

84 An English version is to be found in Scheler (1973). Some other claims related to the topic are present in Scheler (1960) and Scheler (1976) (English: Scheler 2009).

to the kind of organism. The most important feature of human perceptual space is emptiness. Emptiness is constituted by an excess of unsatisfied drives over satisfied ones, something that is in turn a constitutive element of human nature (or, more precisely, mindful creatures in general).

These sentences state only the cornerstones of Scheler's account. I will now explicate some details and relations of these cornerstones.

Scheler does not start his discussion in "Idealism – Realism" with space as it is present for us in perception. Rather, he starts on a more fundamental level with what any kind of space has in common: spatiality. Spatiality is just the bare synchronous being-apart. Spatiality is in effect what makes it possible to distinguish between qualitative identical entities.

Spatiality is grounded in the experienced power to spontaneous movement:

> The central experience [of spatiality] is originally only a creature's experience of his ability to produce certain spontaneous movements. Spatiality is acquired through activity [*erspielt*] before it is represented and thought. (Scheler 1973, p. 330)[85]

In characterizing the movement in question as spontaneous, he highlights that the factual movement that actually happens is not so much of explanatory value as the idea of spontaneity of activity. For Scheler, spontaneity is tightly linked to the idea of a living being: the capability to induce some change to one's state and thereby to elicit another state.[86]

Hence, the first important claim of Scheler's is that spatiality is constituted by this grasped power to move. Spatiality is thus tight to living beings and every living being that is capable of movement and has this basic grasp of his

85 "Das Kernerlebnis ist ursprünglich nur die erlebte *Möglichkeit* oder besser gesagt das *Könnenserlebnis eines Lebewesens*, gewisse *Bewegungen spontan* hervorzubringen. Räumlichkeit wird *erspielt*, *ehe* sie vorgestellt und gedacht wird" (Scheler 1928, p. 296).

86 On Scheler's view, the relevant movement is just a special sort of something more fundamental that all living beings share. This is change. He takes change to lie beyond the distinction of alteration and movement. The latter are special kinds of the former. Alteration (as reversible change) constitutes time, movement (as non-reversible change) constitutes space. For Scheler, time is more fundamental than space. Life is originally a process in time; it spatializes itself; cf. Scheler (1928), pp. 303–4; Scheler (1973), p. 337.

capabilities is thereby capable of experiencing some form of spatiality. Scheler's use of 'experience' ('Erlebnis') of own abilities is certainly not meant in the usual sense as some high-level conscious experience of what one is able to do. I think it is rather meant as some basic grasp of what movements are possible, something that children and animals need to act purposefully.[87]

Scheler claims that every special kind of space, such as our space conceived of as unitary, continuous, three-dimensional, homogenous and unlimited, is relative to the kind of organism (Scheler 1960, p. 303). Accordingly, Scheler explains the properties of space with reference to properties of humans, in particular properties of the movements that humans are capable of (Scheler 1928, pp. 304–5; Scheler 1973, pp. 337–8). All properties of 'our space' are marked as being grounded in properties of the ways we are able to move. Therefore, his statement that "the one subjectively given space is only the sum total of our possibilities of spontaneous movement"[88] (p. 339) has to be taken literally.

One distinctive property of space Scheler explains with reference to something that he takes to be essential to human nature. He claims that only humans (and supposedly sufficiently similar beings, i.e. mindful beings) perceive their space as being void (Scheler 1976, pp. 31–8; Scheler 2009, pp. 26–33). For him, this fact is of "greatest interest" (Scheler 1973, p. 331; Scheler 1928, p. 297). We conceive of space as a substantial emptiness filled with objects. We have the strongest intuition that space does not change when all objects are changed and that it would still be there even if all objects would vanish. Thus, we conceive of the literal non-existent to be more fundamental than the things that properly exist.

This oddity is grounded, according to Scheler, in a constituent of human nature. Humans are creatures where the unsatisfied drives are in excess of the satisfied ones. Humans constantly hold back their instinct or drives and do not satisfy them (Scheler 1928, p. 303; Scheler 1973, p. 336). They do not

87 To catch some prey an animal needs to have some grasp of how it can move, e.g. in which directions it can turn, how far it can reach out or jump etc.

88 "Unsere subjektive Raumgegebenheit [ist] primär nur der Inbegriff unserer spontanen Bewegungsmöglichkeit" (Scheler 1928, p. 306).

blindly follow their instincts but constantly suppress at least part of them. This comes with our being mindful creatures. As Scheler contends, having a mind frees us from the instincts that are related to the living beings that we are, too. Having a mind marks the difference between mere living beings, whose behaviour and perception is dependent on its instincts, and human beings that are able to transcend this sphere. Humans have a world while mere living beings just have an 'Umwelt', as Scheler states, using a central concept from von Uexküll (Scheler 1976, p. 32; Scheler 2009, p. 27). Humans are open to the world. However, humans also still have instincts and these create expectations of what satisfaction of them would mean; but they are never fully satisfied. These unfulfilled expectations ground the phenomenon of the void (Scheler 1928, p. 302; Scheler 1973, p. 335). Other beings that do not have this constant excess of unsatisfied instincts do no experience their space as empty accordingly.

To sum up Scheler's view about the nature of perceptual spatial experience, his account rests on two central claims. First, he claims that spatiality is constituted by the grasped power to self-movement. This surely is the most important claim for my discussion since it formulates the central thesis that action is constitutive of spatial perception. However, it is difficult to pin down a detailed understanding of this point. In particular, it is far from clear what he means by the 'grasped power to self-movement' and, especially, how one should conceive of this grasp. I already said that it is supposedly a low-level mental state representing basic practical capabilities of the subject. The only thing that is clearly indicated by this description is that Scheler's account does not centre on the mere presence of the subject's capabilities (as, e.g. Evans's account does, as we will see) but on some grasp of these capabilities.[89]

The second central claim within Scheler's approach is that the distinctive human perceptual space is further grounded in distinctively human properties. The dimensions of perceptual space are grounded in his ways to move and

89 After having discussed the empirical evidence that speaks in favour of the constitutive views and having argued that on the face of this the capability approaches seem most promising as accounts of perceptual space, I will come back to the question in Ch. 9 of how to develop Scheler's starting points of an account into a comprehensive position.

the fact that we conceive of space as void is explained by a constant excess of unsatisfied instincts over satisfied ones that grounds the void. I think the first of these claims is part of many of the accounts that take action as a constitutive condition on perception. However, the last one is probably the most speculative and queer sounding of the account. I will not elaborate on it further. Note that it already concerns a higher level of representing space than perceiving it. Spatial perception is not necessarily tied to perceiving space as void. It is only tied to perceiving things in a particular distance and directions.

Let me turn towards the question of why Scheler's account might not imply any commitments to embodiment of the perceiving subjects. First, one might again consider the possibility of a bodiless subject and ask whether on Scheler's account it is conceivable that such a creature enjoys spatial experience. I think the situation here is basically the same as in the case of Husserl's account: since space is constituted by the grasped power to self-movement, that is, as we said before with regard to Husserl, by the pure 'I can', no body seems necessary. Hence, if something like a pure *res cogitans* is possible in general, it is also, on this account, likely that it is able to perceive spatially, given it has some grasp of its possibilities to move. In effect, the only thing that is presupposed in Scheler's explanation is a living being, i.e. a being that is able to move actively.

No other conditions need to be satisfied. In particular, the creature does not even need to have means of collecting sensory information. For Scheler, spatiality is constituted independently of and before any sensational information comes in. He adopts a view here that was developed by Erich Jaensch who claimed in his Eidetic Theory that all visual experiences, be it perceptions or imaginations, are dependent on more fundamental eidetic images. Scheler takes this claim as support for his view that spatial experience is constituted by nothing beyond the grasped power to initiate self-movement (Scheler 1928, p. 295; Scheler 1973, p. 329). One might even suppose that what ultimately grounds spatiality in Scheler's account is just the grasp of one's possibilities to move. That is, it is not even necessary that the creature is actually able to move, it just has to have some grasp of himself as having the power to move.

Given that, Scheler's position would be an extreme case of a theory that posits action as a constitutive condition on spatial perception: it is not actual action, it is not past action, it is not actual possibilities of the subject to act but only the possibilities to act that the subject grasps himself to have.

Although these considerations about bodiless beings are clearly just metaphysical speculations, they help to clarify what Scheler's explanation essentially relies on: a being that can grasp himself to have some possibilities to move. Since this is not necessarily connected to the claim about embodied creatures, the explanation does not hinge on any statements about the body.

2.3.3 Gareth Evans

In turning to the position of Gareth Evans, we enter terrain that is more familiar again. The main textual basis is to be found in Evans's incomplete work *The Varieties of Reference* (1982), in particular the sixth chapter about demonstrative identification. Evans holds that demonstrative reference is only possible under two conditions. The first is that there has to be an 'information-link' (e.g. perception) between object and subject; the second is that the subject has to be able to locate the object in space, based on this information-link. (The account is in effect an extension of Peter Strawson's position, in which only the first claim is made; cf. Evans 1982, p. 143.)

In the course of spelling out the details of the second condition, Evans reflects on forms of spatial representation. For him, three conditions have to be satisfied for a given organism to enjoy perceptual space. The first is perceptual input, the second is that this input is directly linked to dispositions to behaviour, and the third is that the organism is capable of conceptual thought.

The last one is a general condition on the possibility of conscious experience on Evans's view. I will not elaborate on this claim but only note that Evans's idea is not that one needs a concept of left and right, e.g., to be able to have conscious experience of space. The connection is more loose in that the subject is supposed to have some thoughts whose (conceptual) content is determined by the contents of his non-conceptual perception. That is, only those contents of perception become conscious that possibly determine the content of (some of) our thoughts.

Evans' first condition on perceptual space is that there is some input from the world to the perceiving subject. Evans calls this 'information' or 'input'. In effect, this is the sensory stimulation that we receive via our sensory systems. Evans claims that the bare proximal stimulation that is caused by the environment is not sufficient for the subject to have spatial experience. Take sound for example. Sound waves, if not approaching from straight ahead, reach the ears with some small difference in time. The waves first hit the ear that is nearer to the source of the sound (interaural time difference). Furthermore, the head in lying between the source of the sound and the second ear shadows the second ear, which leads to a difference in the level of the sound (interaural level difference). These two kinds of differences are important for the lateral discrimination of the direction of sound (left, right, ahead). Others are involved in the remaining directions. The claim that is effectively denied by Evans is that the registration of these differences in the stimulation of our auditory systems alone are already sufficient to elicit spatial sensations (something that arguably holds similar for any sense modality).[90]

That is the point where the second condition comes in. Evans asks the reader to imagine a creature that is able to discriminate between distinctive patterns of proximal stimulation. If one would ask such a creature to respond differently when the left ear first receives a sound or when the right ear first receives a sound, respectively, it would be able to give the right answer. Thus, the organism is able to respond according to interaural differences in time and level of the sound. However, Evans claims that this does not mean *per se* that the creature grasps the direction from which the sound is coming. He claims that only when these patterns of proximal stimulation are directly linked to dispositions to behaviour does the creature experience the sound as having spatial content. Thus, proximal stimulation patterns, although they may vary systematically with spatial content (e.g., a certain interaural time difference correlates to a certain direction of sound), are not in themselves sufficient to

90 He refers to Bower (1974), Rock (1985), Pitcher (1971), and Taylor (1978).

constitute spatial content. Only adding the direct links to action to the picture can explain why we perceive the environment as spatially structured.

Against this line of thought, one might argue that the envisaged creature would indeed be able to know the direction from which the sounds come. One only has to assume that the creature has some background knowledge about the systematic relationship between interaural time and level differences and spatial direction of sounds. That is to say that it could, in principle, calculate the direction of a particular sound, given that he knows the actual values of the relevant differences and some equation that specifies the relation between the pattern of differences and the angle of the approaching sound.

Evans's answer to this objection could be that it is unlikely to be applied to the human case. Given that it is true, it would follow that it is possible that two people hearing a sound from the same direction could react differently when they would be asked, e.g., to orient towards the sound. This is due to the possibility that there are mistakes or simply differences in their calculations. Since this never happens (or so claims Evans), we have good reason to suppose that the spatial content is not the result of some calculation that grounds our reaction. Rather, it is more likely that it is the other way around; the dispositions to behaviour are immediately linked to the input pattern of proximal stimulation, and are thus co-constitutive of spatial experience. Therefore, individual differences are ruled out; since a given location is constitutively dependent on some disposition to behaviour, two people experiencing a sound as coming from the same direction cannot but react the same way, since the disposition to exactly this reaction itself constitutes the kind of spatial experience they share.

Now, one might again argue against this that individual differences in reaction to sound are not common because the calculations are not that complex or they are hard-wired into our cognitive machinery. This would be an alternative explanation of the absence of differences in reactive behaviour. Another way to argue against Evans would be to say that it is not obvious that there are no such differences; maybe they are quite small and difficult to measure so that they are not observable in daily life. Evans simply assumes that there are no such differences in behavioural reactions, but he does not show that this is actually true.

However, this point can be answered as well. It seems to me that Evans's point here is not to make an inductive inference from the observation that there are no behavioural differences to the general claim that dispositions to behaviour are constitutive of spatial perception. That would be an unwarranted move. It is rather that the absence of behavioural differences follows from his account easily. Therefore, one should read it more like something to raise the plausibility of the account. I think Evans's fundamental point is that it is in principle impossible to get auditory spatial perception from the registration of interaural time and level differences alone. One could imagine a creature that does this registering for a hundred million years; it would at most end up with a plethora of possible time and level differences and maybe with knowledge about some systematic relations between them. However, it could never be able to assign these differences some spatial meaning (Grush 2007, p. 406). How should it? It just registers that there are these differences in the stimulation of its ears. Before, I said it would be possible for such a creature to calculate the direction if it had enough knowledge about correlations between the values of the interaural differences and differences in direction of sound. However, the exact question is how the creature should ever know about these correlations by only being able to register the differences. This is the point where the dispositions to act come in. The sounds do have to be directly and systematically linked to dispositions to behaviour. These dispositions are in fact what one might regard as a way of knowing the correlations. The correlations between interaural differences and direction of sound are somehow present in these dispositions. The dispositions that are connected with particular proximal stimulation patterns therefore are what bridges the gap between values of interaural differences and genuine spatial sensation. Thus, "only agents are perceivers" (Schellenberg 2007, p. 603).[91]

91 Peacocke (1983), Ch. 3 develops an account of spatial representation that is inspired by and an extension of Evans's approach. On Peacocke's account, spatial representation is constituted by links between behaviour and sensation. He argues that not just any input-

Now, let me turn to the question of how Evans's position differs from the embodiment approaches I discussed in the preceding chapter. One of the properties of perceptual space is that it is organized around a centre. Usually the centre of my perceptual space is where I am, that is, where my body is. I perceive things as being located in relation to the position of my body. This is natural on Evans's picture since I perceive with my bodily sensory systems and the behaviour, the dispositions to which are linked to the proximal stimulation, is usually bodily behaviour. Thus, I do not only perceive from where my body is but I also act from where my body is. The body seems to be the natural centre around which perceptual space is structured. What I regard as left and right is actually defined in relation to the position of my body, and it changes with this position potentially. Hence, it seems clear that my body is also partly constitutive of the meaningful structure of my perceptual space. The body defines the place we would call 'here'.

However, we find a possible scenario in Evans's work where it is not necessarily the case that the centre of egocentric space is defined by where my body actually is. In order to reveal a detailed understanding of how 'here'-thoughts work, Evans considers a case in which someone remotely controls a submarine from a vessel on the surface of the sea (Evans 1982, pp. 165–7). The submarine is equipped with cameras, excavators, 'limbs' and propulsion mechanisms that are controlled by the subject. The subject is highly trained in controlling the submarine. Since he sees nothing other than the pictures delivered by the cameras, has no other perception about what is going on in the vessel and is able to fluently 'act' with the submarine, Evans regards it as

output connection is sufficient for constituting spatial contents. For example, mere stimulus-response connections are not enough, since they are in principle present below the level of content. Furthermore, Peacocke emphasizes that a representation of just one place is not possible and that every individual capable of spatial representation must at least being capable of representing two places. This means that spatial representation necessarily implies some kind of mental map of the represented places. Thus, Peacocke refines the relevant notion behaviour that is taken as constitutive and he puts stronger conditions on representing egocentric space. Apart from these distinguishing features, Peacocke's account can be grouped with the capability approaches. Finally, one issue with Peacocke's account is that he shifts between giving conditions on spatial representation and conditions on the applicability of spatial representations. Arguably, both do not have to coincide.

possible that the subject may become so immersed into the scenery at the sea bed, that he finally regards himself to be at the place where the submarine actually is. In this case he would be able to think of the objects he perceives at the ground of the sea as 'this X' just like we normally refer to things and places in our environment demonstratively. This would also mean that his 'here' thoughts refer to the place where the submarine is and not where the body of the subject is, namely in the vessel at the surface.

One might object to this scenario that there will always be some awareness on the side of the subject of the fact that it is not really situated at the ground of the sea, that it only pretends to be there. The situation is too strange. In addition, it is likely that there are time lapses between reactions of the subject and movements of the submarine etc. However, it seems to me that this scenario is actually possible. A similar and perhaps more intuitively appealing one would be someone playing a video game, in particular, one of that kind where one perceives and interacts with the environment from a first-person perspective. It seems very likely to suppose that after playing some time and being really used to the controls and everything it is possible that one gets so immersed in the game that you not only would refer to the things you see there the same way as when you refer to things in the real world. Furthermore, everything you perceive you would perceive in relation to the place in the virtual environment. That means that you would locate e.g. sounds you hear that are in the real world in the virtual world. You would turn your virtual avatar, not yourself, in reacting to a sound. To refer to the actual sound in your real environment you would then have to think of the sound as 'coming from my living room', i.e. you would not refer demonstratively to the sound in the usual simple way.

If such cases are possible, something that is very likely in my impression, it is also possible that the centre of my egocentric space, the 'here' is dissociated from the place where my body actually is. In that sense, Evans's theory does not rely on claims about the body. Instead it allows that the centre of egocentric space is not defined by the (position of) the body but by the 'origin of the movements that one is disposed to'. The reason why it seems initially that the body determines the centre would then just be contingent on the fact that in normal circumstances our body is exactly there where this 'point of origin of our actions' is.

The general idea might be expressed by stating that the centre of egocentric space is not where your body is but where you conceive yourself to be. This in turn depends on the practical possibilities you conceive of yourself as having to change and interact with what you perceive. It is true that, in one sense, in the described scenarios, you act by e.g. pressing some buttons with your actual body. However, if you conceive of your possibilities as possibilities of pressing buttons in a certain way, you are not really immersed in the situation. Thus, someone who would object to the presented view in saying that the dispositions you have are really only dispositions to press buttons, would deny the conceivability of the scenario altogether. However, part of the scenarios is that the button pressing is not noticed in any way anymore. It becomes just another step in the control of behaviour as the steps that are beyond your awareness in the usual case, like bodily processes that make up your movements. Instead, given the scenario, you would describe your possibilities as turning this or that way or moving left or ahead, thereby expressing possibilities you have in being at the seabed or in the virtual environment. Space is centred on where you act not where you really are on Evans's account.

2.3.4 Summary of capability accounts

The presentation of the three accounts has revealed that all of them share the basic conviction that the capability to act is essential for an explanation of spatial experience. However, there are differences in the way this idea is cashed out. Husserl's account is a bit difficult to spell out due to the unsystematic presentation. However, I think, it is in some sense more akin to Evans's than to Scheler's. He emphasizes the possibilities that the subject actually has and that the centre of the perceptual space is defined by these possibilities. If my interpretation of Evans's view is apt, these two claims are also part of his account. In contrast, Scheler's account crucially differs with respect to the other two in that he sees some basic sense or grasp of one's own practical possibilities as constitutive. The mere presence of the latter is not enough according to him.

I also made clear in what sense these accounts differ from the embodiment approaches I presented before. I see in each account hints to the effect that the sense in which action is constitutive for them does not depend on claims about embodiment. They may well be compatible with such accounts but, importantly, the claims made here are independent of any prior commitments about the embodied nature of subjects. This is what distinguishes them from the other group of accounts, exemplified by Gallagher and Merleau-Ponty above.

2.3.5 Summary of systematic overview

In the last two Chapters, I have presented the details of four kinds of accounts claiming that action is a constitutive condition on spatial perception. Each of the accounts takes a different dimension of action to be the crucial. Hurley's explanation operates largely on the level of sub-personal processing and the proposed mutual relationships between mechanisms that subserve perception and mechanisms that subserve action. Her basic distinctions, especially the one between instrumental and non-instrumental dependence are illuminating and helped to structure the extent accounts. However, I also argued that her account lacks the generality that is needed for my general argument to the effect that egocentric spatial perception is a kind of self-representation, a generality that is endorsed by the remaining accounts. Some more problems regarding Hurley's account will come up in section 3.1.2, where empirical evidence will be discussed in light of her view.

Alva Noë's approach puts the emphasis on a distinctive kind of practical knowledge. Unfortunately, the exact shape of his account remains somewhat unclear due to the difficulties with interpreting his writings. What seems to be clear, though, is that sensorimotor knowledge, the crucial element in his account, is gathered by self-induced movements. Thus, he claims a rather indirect connection between action and spatial perception compared to the other accounts. In the next chapter, I will present empirical evidence that speaks against the sensorimotor view (see section 3.1.3 and 3.1.4 below).

The remaining approaches give an account of spatial perception in terms of the capability to act. I distinguished two sub-groups of these accounts depending on the importance that notions associated with embodiment play in the accounts. Shaun Gallagher's and Maurice Merleau-Ponty's views are explicitly formulated in terms of the basic automatic interactions that we are capable of by being the embodied creatures that we are. We saw that Gallagher's account is particularly hard to understand. In Merleau-Ponty's writings, there are also some notions, e.g. 'gearing', that remain unclear to a certain extent. A common mark of these approaches is a certain anti-intellectualistic attitude towards perception: everything necessary for spatial perception to be present is already there in the supposedly phylogenetically ancient low-level bodily systems that enable the organism to cope with the world.

This attitude stands in a quite sharp contrast to a basic tenet of Max Scheler's approach. Although, on his account, perceptual space also is tied to some very basic capability, namely to a grasp of one's own action possibilities, some features of perceptual space are explained by distinctive human properties like the constant excess of unsatisfied drives that constitutes the void. The fact that it is a basic sense about what is possible for one and not the mere presence of these possibilities that constitutes spatial experience highlights a more sophisticated structure than is supposed by Gallagher or Merleau-Ponty.

Finally, I outlined Gareth Evans's account as the one that is present in the background of many rather recent discussions of the issue (e.g. Campbell 1993; 1994; Schellenberg 2007; Grush 2000; 2007; Briscoe 2009). I outlined his basic idea that informational input has to be directly linked to dispositions to act to have any spatial significance for the subject. His approach does not rely on some basic grasp or sense of one's practical possibilities as Scheler's does, which is the crucial difference between the two.

In the following chapter, I will discuss different kinds of empirical evidence that are relevant to the topic. This shall not only further add to the substance of the claim that action is a constitutive condition on spatial perception; it shall also help to decide which of the accounts is most appropriate. All kinds of approaches will feature again in the discussion in some way.

3 Empirical Evidence

In Ch. 2.1.2, I discussed the notion of constitutive dependence and ended by endorsing Burge's view. On his account, constitutive explanation concerns the nature, or essence, of a phenomenon. Part of it is the claim that constitutive questions can and often have to be answered by drawing on empirical evidence.[92] I think this is right and, hence, in this part, I will discuss empirical evidence from various fields that shall help to answer the questions whether and in what way action is a constitutive condition on spatial perception.

In the preceding two chapters, I outlined the major accounts that defend a constitutive dependence of spatial perception on action. However, I have not presented compelling reasons for adopting one or the other view yet. Thus, the empirical evidence that I will discuss in Ch. 3.1 shall serve two purposes. On the one hand, discussing this material adds plausibility to the general claim that a constitutive explanation of spatial perception needs to take action into account. On the other hand, some results of the discussed studies can motivate a decision about which of the presented accounts is most promising. In particular, I will discuss experimental studies that reveal effects that actions have on spatial perceptions, studies on inverted vision and studies on sensory substitution devices.

The picture that will emerge is quite complex. All kinds of evidence have been cited in the literature as straightforwardly speaking in favour of the constitutive dependence of perception on action. Indeed, I will also argue that, in sum, these cases present good evidence. However, my discussion will show that the situation is rather complex and that the different kinds of evidence speak in favour of different versions of the constitutive claim. Thus, for each kind of evidence, I will discuss in what way the results are indicative of a

92 "[A]nswers to constitutive questions [...] often have empirical warrant" (Burge 2010, p. 59).

constitutive dependence of spatial perception on action. I will argue that in light of the empirical results only the capability approaches can be sustained.

In Ch. 3.2, I will consider an empirical challenge for the constitutive approach. In particular, it has been argued that the dual streams approach to vision presents a challenge to theories that posit a constitutive dependence of perception on action. I will discuss the challenge and argue that it is actually no threat for accounts that posit a constitutive dependence between perception and action.

In a short Appendix to this chapter, I want to discuss various philosophical arguments that centre on the dual visual streams approach. I will largely argue that the arguments are mistaken and should therefore be abandoned.

3.1 Empirical evidence for action as a constitutive condition on spatial perception

In this chapter, I will discuss three sorts of empirical evidence that support the claim that action is a constitutive condition on spatial perception. This shall serve as general evidence for the claim that spatial perception depends constitutively on action, a claim that was hitherto investigated only systematically and historically. Furthermore, this evidence will be a means to decide between the different accounts discussed in the last part.

Before outlining the chapter, let me note two things. First, it is not common to speak of action as a 'constitutive condition' on spatial perception in the empirical literature to be discussed. Instead, researchers typically write of action having an 'effect' on perception. Therefore, part of my task will be to explore whether such claims might be interpreted more robustly, i.e. as pertaining to constitutive conditions on perception.[93] It might well be that action has great effects on perception without it being a constitutive condition on

93 Cf. e.g. the work of Dennis Proffitt's lab discussed below.

perception. This was made clear in the discussion of the difference between what Hurley dubbed 'instrumental dependence' vs. constitutive dependence (see Ch. 2.2.2 above).

A final preliminary remark concerns the term 'action'. I already pointed out different ways in which action figures in different accounts that take action as a constitutive condition on spatial perception in the systematic overview of accounts. It will become clear that in the empirical literature on the matter the same holds true. Researchers focus on very different features of action in defending the claim that action is a constitutive condition on spatial perception. The proposals range from actual movements to planned and/or imagined movements to anticipations of effort that is associated with certain acts to sub-personal processes such as motor simulation. To understand the claims and their significance for the constitutive claim, it is important to figure out the feature of action that is central to each account.

Let me start with an overview of the empirical evidence that I want to discuss in this chapter.

3.1.1 Overview of empirical evidence in favour of constitutive views

Figure 5 provides a table that lists the three kinds of evidence that I will tackle in this chapter:

OVERVIEW OF EMPIRICAL EVIDENCE

	'Proffitt studies' (Ch. 7b)	Inverted vision studies (Ch. 7c)	Sensory subst. devices (Ch. 7d)
EFFECTS	– perception of distance, slant	– orientation of visual field	– presence of perceptual space
INDUCED BY	– tool-use, wearing a back-pack, walking on a tread-mill, intention to throw a ball	– wearing inverting goggles	– auditory/tactile substitution devices for vision
SUPPORTED CLAIM	Particular actions are a constitutive condition on particular perceptions.	Ability to act is a constitutive condition on particular perceptions.	Ability to act is a constitutive condition on ability of spatial perception.

Figure 5: Overview of the sources of empirical evidence and the central claim based on them.

My discussion starts with the 'Proffitt studies' (most of them are from Dennis Proffitt and his colleagues, therefore the label). These studies centre on action-induced changes of particular dimensions of perceptual space, like egocentric distance, slant of a hill, height of an abyss, or size of an object. I will discuss various studies that use different measures and propose different explicative hypotheses that are related to different aspects of action.[94] I will also discuss some criticisms that have been put forward against the validity of the results.[95] Finally, the question that is most important in this context is what relation of action to spatial perception is revealed in these experiments. I will draw on Hurley's contrast between instrumental and constitutive dependence in tackling this question. Since the results of these studies are indicative of

94 Here is a list of examples, though I will discuss not all of them: Berti and Frassinetti (2000); Farnè and Làdavas (2000); Iriki et al. (1996); Kirsch and Kunde (2013); Làdavas and Serino (2008); Linkenauger et al. (2006); Longo and Lourenco (2006); Osiurak et al. (2012); Proffitt (2006); Proffitt (2008); Proffitt and Linkenauger (2013); Proffitt et al. (2003); Rizzolatti et al. (1997); Schnall et al. (2008); Schnall et al. (2010); Stefanucci et al. (2008); van der Hoort et al. (2011); Witt (2011); Witt and Proffitt (2008); Witt et al. (2004); Witt et al. (2005).

95 E.g. Corlett et al. (1990); Durgin et al. (2009); Durgin et al. (2012); Firestone (2013); Firestone and Scholl (2014); Grave et al. (2011); Hutchison and Loomis (2006); Russell and Durgin (2008); Woods et al. (2009).

effects that particular actions / intentions have on particular episodes of spatial perception, they can at most serve as evidence for Hurley's account of the constitutive dependence. Hers was the one I discussed that is explicitly formulated in terms of particular instances of action and perception. However, my discussion will reveal an ambiguous picture. I will show that it might be possible that these cases exemplify merely an instrumental dependence. Finally assessing the cases would require additional empirical work.

The second group of experiments that I will discuss in section 3.1.3 involve studies on inverted vision.[96] Since Stratton did the first experiments using inverted vision, there is an ongoing discussion about the right interpretation of the results and the right way to describe the phenomenology. The results of Stratton have been replicated e.g. by Kohler, Dolezal, and most recently by Sekiyama and colleagues. All these researchers claim that after a certain time of being exposed to an inverted visual field, normal, upright, vision is re-instantiated. However, there is also competing evidence, e.g. Linden et al. (1999), where no such effect is observed. I will argue that, overall, the evidence indicates that there *is* such a re-instantiation of upright vision. Therefore, inverted vision studies can be used in arguing for the constitutive claim. My discussion will centre largely on a certain stage, which is described in all reports and which seems problematic for constitutive views. During the experiments, which usually last for several days, there is a phase where subjects are able to interact with objects they perceive in a seemingly normal and fluent way, while their perceptual field remains inverted. On the face of it, this seems problematic for constitutive views since one would expect the perceptual field to re-invert *as soon as* the capability to act based on what one sees is present. I will argue that this observation presents defeating evidence against the sensorimotor approach and I will consider different ways in which capability approaches can accommodate this observation.

Finally, in section 3.1.4, I will discuss studies involving sensory substitution devices. These devices are used by the blind to regain (some aspects of)

96 Stratton (1897); Kohler (1951; 1964); Taylor (1962); Dolezal (1982); Sekiyama et al. (2012); Degenaar (2014).

their lost vision. These studies are most fundamental since they show the establishment of spatial content from non-spatial sensations (e.g., sound patterns enable subjects to locate objects in their vicinity). However, before assessing the significance of these cases for the constitutive views, a problem needs to be resolved. In particular, it has to be shown first that these cases are genuine cases of spatial perception and that the results cannot be explained by judgments based on auditory or tactile perception. I think that this is a major issue for using these cases as evidence, which has not received enough attention yet. I will argue that there are good reasons to suppose that there actually is spatial perception present in these cases after sufficient training. Finally, I will ask which account best suits the experimental results and argue that a version of the dispositional view seems best while these cases are highly problematic for the sensorimotor approach.

The three kinds of studies differ in two respects. On the one hand, the studies differ in the spatial dimensions of visual perceptual information that is altered. Dimensions in this sense are direction, distance, orientation (slant) and size. Whereas the Proffitt studies centre on perceptions of distance, size or orientation, the inverted vision studies centre on the aspect of direction. The investigations on sensory substitution devices centre on the generation of spatial significance of non-spatial sensations, i.e. how non-spatial information acquires direction and/or distance information. I take it that the third source of evidence provides the most powerful argument for the constitutive view since in these cases there is not only an alteration of certain aspects of spatial perception but a transition from non-spatial sensations to spatial perceptions although input remains fixed. Therefore, these cases allow a comparison between a subject that enjoys spatial perception and one that does not although both receive the same information in terms of proximal stimulation.

The second respect in which the discussed kinds of evidence differ consists in the *relata* that are claimed to be dependent on each other. The Proffitt studies are supposed to show effects that *particular* actions have on *particular* perceptual states. The inverted vision experiments show how the *capability* to act affects the spatial structure of perceptions. Finally, the studies on sensory substitution devices are supposed to show how the *ability* to perceive spatially is dependent on the *ability* to act. These differences might be subtle but they are important to keep in mind since they effectively lead to different versions of the constitutive claim. Furthermore, they do not simply entail each other. Although it is true that capability claims can be translated into claims about particular cases, the opposite is not true; to show that in certain cases action has an effect on spatial perception is not sufficient to show that the ability for spatial perception depends on action.

In general, my discussion of every group of studies will proceed in three steps. The first is the presentation of the setting and main results of the studies. The second will consist in dealing with questions of validity and appropriate interpretation of the results. Some studies face criticisms that question the validity of the results, others are criticised based on methodological issues and for the final two sources of evidence in particular, there are disputes about how to best describe and interpret the experiences that are made. It will therefore be an important task to assess as good as possible what can be taken as empirically established facts and what cannot. This relates both to observed effects and proposed explanations. The third step is to assess the relevance of the results with regard to the claim that action is a constitutive condition on spatial representation. The third is the more important and interesting task. However, it is pointless and redundant without having accomplished the second.

3.1.2 Effects of action on distance and slant perception – The Proffitt studies

Several studies aim at confirming the hypothesis that action has an effect on perceived distance and slant. The most prominent ones come from Dennis Proffitt's lab. The most interesting experiments for our purposes are those that show effects on distance or slant perception induced by tool-use, wearing a backpack, walking on a treadmill, or intending to throw a ball, respectively. In all cases, it is the performance of or the intention to a particular action that has an effect on spatial perception. Thus, *prima facie*, these experiments straightforwardly reveal a constitutive dependence of perception on action. I will first present the central studies and then discuss in what way they might indicate a constitutive dependence of spatial perception on action. Since these cases centre on particular actions / perceptions they are at most evidence for the constitutive claim as developed by Hurley.

Let me start with a brief description of the relevant sorts of studies.

The first kind of study I want to discuss comprises experiments that investigate effects of tool-use on perceived distance. Witt et al. (2005) reports several experiments that are supposed to show that usage of a tool can change perceived distance. They found that participants perceive a target to be closer in case they intend to reach it with a tool compared to when they intend to point at it. Critically, the target in both conditions is at the same actual distance. Participants who are intending to use a tool to reach the target perceive it about seven centimetres closer as participants reaching only with their finger (the actual distance of the target varied between 70 cm and 120 cm).

In the same study, Witt et al. show that the *intention* to use the tool is critical: in their third experiment, one group of participants merely held the tool but did not intend to reach with it. In this case, no difference in perceived distance between tool-users and those that point with their finger was observed. Witt and Proffitt (2008) suggest that even *imagining* reaching with a tool affects perceived distance. Furthermore, the size of the tool makes a difference. In an experiment reported in Osiurak et al. (2012), participants that held a tool that was actually too short to reach the target showed no effect on perceived distance.

The authors claim based on the results that the subject's estimation of how far it can reach affects distance perception. Witt and Proffitt (2008) propose motor simulation as an underlying mechanism for this effect.[97] Thus, they argue that the outcome of a motor simulation, which takes into account whether a tool is used or not, affects perception of a target's distance. This is

97 Their account is based on the notion of motor simulation that is developed by Jeannerod (2001); cf. for a similar account Kirsch and Kunde (2013).

why the effect is even observed in cases of imagination and not observed when the tool is held without any intention to reach with it.

Another study that received great attention in the literature measured the impact of wearing a backpack on the perception of slant (Bhalla and Proffitt 1999). The results of this study suggest that wearing a heavy backpack lets hills look steeper compared to when not wearing a backpack.[98] Bhalla and Proffitt report that participants who did wear a backpack estimated the slant being 27% steeper compared to participants wearing no backpack (as measured by a visual matching task). The authors attribute the effects on perception to altered "physiological potential". In another paper, they interpret the results as being grounded in the alteration of anticipated effort that is associated with climbing the hill (Witt and Proffitt 2008, p. 1488).

A change in anticipated effort is also supposed to explain altered estimates of distance (Proffitt et al. 2003; Witt et al. 2004). In experiment 1 of Proffitt et al. (2003), it is reported that distances are perceived as greater by participants that wear a backpack compared to a control group that did not.[99]

The other two experiments reported in that paper use other means to argue for the effect of anticipated effort on distance perception. In these experiments, walking effort was changed by means of a treadmill. One group of participants walked on a treadmill while perceiving no optic flow (using VR), whereas another group of participants perceived optic flow while walking on the treadmill. The effect of this task is that the anticipated effort of staying at the same place is recalibrated. Thus, when the participants that received no optic flow after completing the treadmill task were told to walk in place while blindfolded they tended to walk forward. This aftereffect is significantly greater for the group that received no optic flow than for the control group that received optic flow while walking on the treadmill. Now, based on the

98 It is generally the case that slants are highly overestimated by participants. Thus, even participants that do not wear a backpack estimate a slant of 5° to be of about 20°.

99 Strictly speaking, participants wearing a backpack underestimated the distances not as much as participants that did not wear anything. Thus, as in the case of slant perception, people are generally bad at estimating distance; in particular, it is commonly underestimated; cf. references in Proffitt et al. (2003), p. 107.

hypothesis that anticipated effort affects distance perception, the authors predicted that the group that did not perceive any optic flow would perceive a distance to be greater than the control group, since they would associate a higher amount of effort with walking to a given target. The rationale for this idea is that since staying in the same place requires some effort already, walking to a distant place should be associated with increased effort compared to the effort that those participants associate with walking to a distant place that experienced no effort when staying in the same place. In fact, as predicted, the results indicate that those participants that received no optic flow estimated the distances to be greater than the control group.

In another study that investigated the role of anticipated effort, Witt et al. (2004) used the same treadmill paradigm to further investigate the role that intentions play. As mentioned before, Witt et al. (2005) proposed that the intention to use a tool is critical. In a similar way, the authors in their study from 2004 propose that changing the effort associated with walking via the means of the treadmill and elimination of optic flow only has an effect on distance perception in case the distance is perceived while intending to walk to the target (cf. Rieser et al. 1995). Thus, participants who intended to throw a ball to the target after estimating the distance showed no effect of the treadmill-induced recalibration on perceived distance. The authors conclude that anticipated effort affects distance perception not *simpliciter*, but that the effects are "functionally specific". This means that distance perception is affected by the distinctive amount of effort one associates with the particular action one intends while perceiving the distance. Thus, the anticipated effort of walking to the target does not affect perception of the distance of the target when intending to throw to the target, and vice versa.

I take it that the studies I presented are enough to illustrate the kind of investigation that the members of Proffitt's lab in particular undertake. I did not mention here experiments that investigate effects on size perception (e.g. Witt and Dorsch 2009; Witt et al. 2008; Witt and Proffitt 2005), effects of emotions on perception of height (Stefanucci et al. 2008), or effects of taking glucose on slant perception (Schnall et al. 2010).

What explanation do Proffitt et al. offer for the observed effects? These studies are supposed to show that spatial perception cannot be explained exhaustively by the optical information that is transduced by the eyes. The authors claim, "[w]e see the world in terms of the potential actions that it affords and the effort associated with these actions" (Witt et al. 2004, p. 590). On other occasions, they speak about the influence that "physical potential" (Bhalla and Proffitt 1999, p. 1093), "energetic costs" (Proffitt 2006), or "abilities" (Proffitt 2008; Witt 2011) have on spatial perception. Although these notions admittedly differ, they point in the same direction.

I think that Witt and Proffitt (2008) develop the clearest version of the account. Therefore, I will rely on this paper in discussing the implications of their view. In this paper, the authors do not only state that spatial perception is a "function of optical information, the perceiver's intention, and the perceiver's ability to perform an intended action" (p. 1490), thus reiterating some of the elements already cited above, but they also present some evidence about the nature of (at least one of) the underlying mechanisms.

They argue that three factors determine spatial perception: (1) optical information, i.e. patterns of light on the retina that gets processed in the visual cortices; (2) occurrent intentions the subject has as he perceives the environment; (3) some assessment of the ability the subject has for performing the intention, including a 'consideration' about available tools and an anticipation of required effort to perform the intended action.[100] The second and third of these factors play a role in motor simulation, which is claimed to be the underlying mechanism of the observed effects.[101]

The notion of 'motor simulation' is taken from Jeannerod (2001). According to him, it is "a representation of the future, which includes the goal of the action, the means to reach it, and its consequences on the organism and the external world" (p. S103). Motor simulation is commonly taken to reflect the

100 Factor (3) is not taken to involve any high-level cognitive processes such as deliberating. I take it that it is supposed to be some intuitive grasp of the ability and the relevant facts.

101 The authors aim to establish this point by conducting experiments that are supposed to distract the participants from entertaining the motor simulation, which results in an altered pattern of results compared to participants that are not distracted.

constraints under which actual bodily actions are performed. Thus, as Jeannerod claims, motor simulations essentially share all features of actual bodily actions except for one: the execution (cf. Witt and Proffitt 2008). This renders them as especially promising candidates for how the effects observed by Proffitt et al. are realized in the brain.

The explication of a plausible candidate for a mechanism that underlies the supposed effects surely strengthens the general account of 'action-specific perception' endorsed by Proffitt and his colleagues. However, there are also a number of problems related to these studies. In the following section, I will review empirical results that question the results of Proffitt et al. Subsequently, I will present alternative hypotheses that may explain the findings. Together, these issues present some empirical problems of how to assess the evidence of the 'action-specific approach to perception'. After having presented these issues, I will go on to discuss what kind of relation between action and spatial perception this body of evidence, if any, reveals.

3.1.2.1 Empirical objections to Proffitt studies

First, let me note that some of Proffitt et al.'s results could not be replicated. I will cite these cases here but I think they do not undermine the entire approach. De Grave et al. (2011) (a replication of Witt et al. 2005) found no effect of intending to use a tool on distance perception. Furthermore, Durgin et al. (2009) investigated the supposed effect of wearing a backpack on slant perception (as originally reported in Bhalla and Proffitt 1999). Durgin et al. had three groups of participants. The participants of two groups wore a backpack. Participants of the one group were told a "cover story" that warranted the necessity of wearing the backpack.[102] Interestingly, the backpack did only have an effect on slant perception for the group that wore the backpack *without* receiving an explanation for the need of the backpack. The participants that received the explanation for wearing the backpack, however, showed the

102 Participants in this group were told that the backpack contained some equipment that was used to monitor their ankle muscles. They also got some electrodes put to their ankles and there was a fan inside the backpack to increase the credibility of this cover story for why the backpack would be necessary.

same results as the control group, i.e. those participants that did not wear a backpack at all. Furthermore, their results indicate that those participants that believed that the backpack was supposed to alter the perception of the slant showed an even greater effect on slant perception.[103]

Finally, Woods et al. (2009) found that effects of ball weight on distance estimates are conditional on the way participants are instructed to estimate the distance. Only those participants who were explicitly instructed to take into account nonvisual factors in estimating showed an effect. The results for this group were the same as those reported in Witt et al. (2004).[104]

Durgin, Woods and their colleagues propose that *demand characteristics* explain the results of the Proffitt studies. The term 'demand characteristics' refers to cues that let participants guess the point of an experiment (Orne 2009; 1962). These can comprise the experimental setup, the explicit instructions given to the participants, or the experimental procedure itself etc. In general, it is of utmost importance to prevent participants from guessing the experimental hypothesis that the researchers try to confirm since this guess might influence the performance of the participants.

103 Durgin et al. (2012) report another study that is supposed to question the results of Bhalla and Proffitt (1999). In this study, all participants wore the backpack. There were two groups that received different instructions concerning the way they should estimate the slant. One group was told to ignore the backpack while estimating. Both groups showed significantly different results, with the 'ignore!'-group displaying the same pattern of results than a control group that did not wear a backpack. The authors conclude that this study also shows that wearing a backpack does not in general has an effect on slant perception. However, I think there is a problem with this paradigm. As we saw before, Proffitt et al. maintain that anticipated effort that is somehow computed by a motor simulation is responsible for the effects. They even found that imagining e.g. holding a tool can have an effect. Hence, a possible answer to this study available for them would be that those participants that should ignore the backpack were able to simulate the effort that was necessary in case they would not wear a backpack and therefore they showed a different pattern of results. In fact, this would be consistent with Proffitt et al.'s general approach. Therefore, I regard this latter study from Durgin et al. not as competing evidence against the results from Proffitt/Bhalla in particular.

104 In contrast, participants of the other two groups that should report "apparent (perceived) distance" or "physical (objective) distance", respectively, did not show any effect of the weight of the thrown ball.

Durgin et al. and Woods et al. claim that some of Proffitt's results are only changes in judgments elicited by demand characteristics. In effect, on their view, the changes in distance or slant estimation are neither perceptual nor an action effect but simply due to participant's attempts to confirm the guessed hypotheses. This would also explain their own results that distance and slant estimation change with a difference in the understanding of distance (Woods et al. 2009), or depending on whether an explanation for carrying the backpack is presented (Durgin et al. 2009).

The crucial question is whether this criticism applies to all of the experiments that form the backbone of the action-specific approach to spatial perception. However, the observations from Durgin et al. and Woods et al. only pertain to two experimental paradigms of Proffitt's and they have not been replicated yet.

Hence, I think that Woods et al. (2009) are right in being cautious about drawing too strong conclusions from their own results. In particular, they do not want to claim that their objections are straightforwardly generalizable to the entirety of experimental paradigms. They also do not claim that Proffitt et al.'s action-specific approach *per se* is misguided but merely that not all experiments that are cited in its favour are expressive of the proposed effects. They argue that, with regard to the current standard of knowledge, action might affect perception in the way Proffitt et al. propose, however, it depends on the context whether this is the case.

In my view, this is the most reasonable attitude to hold with regard to the Proffitt studies. I think that especially those experimental paradigms that include changes of optic flow can support the action-specific hypothesis since in these experiments it seems to me much less likely (compared to the other studies) that participants would guess the significance of the experimental setup. In addition, the results of these studies have been replicated several times (Rieser et al. 1995; Proffitt et al. 2003; Witt et al. 2004). Therefore, I think not all studies are threatened by the issue of demand characteristics.

To sum up this point, we saw that there are some methodological issues concerning the experimental design of the Proffitt studies. However, in what follows, I will take it for granted that at least some of the cases that Proffitt and others present are actual cases of action affecting spatial perception.

In the following section, I want to discuss what kind of dependence of spatial perception on action is supported by these cases.

3.1.2.2 What kind of dependence?

The experimental design of the Proffitt studies concerns the effects that particular actions have on particular spatial perceptions. Therefore, they can only reveal a constitutive dependence of spatial perception on action in Hurley's sense. The other approaches claim that the *ability* to act is a constitutive condition. Of the accounts that I discussed in the last two chapters, only Hurley's concerns the constitutive conditions of particular perceptions. Therefore, the question arises whether the Proffitt cases are examples of a constitutive dependence in Hurley's sense.

I will discuss whether the results of the Proffitt studies indicate a constitutive dependence of spatial perception on action in Hurley's sense. I will argue that the results are in fact ambiguous. This is due to a difficulty in understanding Hurley's idea of an *internal* instrumental dependence. Finally, I will consider the significance of Hurley's notion of constitutive dependence with regard to my overall argument.

Let me briefly remind you of Hurley's account of the constitutive dependence of spatial perception on action (see Ch. 5.2.2 above). She contrasts the notion of constitutive dependence with the notion of instrumental dependence (Hurley 1998; Hurley 2001). The core idea of the instrumental dependence between spatial perception and action is that perception is dependent on acts in the sense that these acts change the causal relation between environment and sensory systems. To perceive what is behind some obstacle one has to move a certain way. That is, "[t]he dependence of the contents of perception on output is instrumental when it operates via input rather than directly" (Hurley 1998, p. 362). Perception is instrumentally dependent on action, and in particular movement, in the sense that movement is necessary for the subject to get into contact with certain things in the environment that are not perceived without movement. Movement thus changes the input pattern, e.g.

the pattern of light on the retina, which in turn results in altered perception. Therefore, movement is a means to perceive certain things that are otherwise not perceivable. The kind of movement in question is not restricted to active explorations of the subject. In principle, the same effects of changes of retinal pattern may be accomplished by mere passive movements or movement of the environment. This is also a mark of the instrumental dependence.

Do the results from the Proffitt studies indicate an instrumental dependence? It seems not since no actual movement is supposed to account for the effects, hence, no altered causal configuration between eyes and environment. What differs between the experimental conditions are internal states like intentions, or anticipations of effort, and body-related changes such as holding a tool or wearing a backpack. The one element that is essential to the instrumental dependence, namely movement, does not play any relevant role at all in these experiments.[105]

This suggests that the observed effects are expressive of a constitutive dependence of spatial perception on action. Here is again Hurley's characterization of the constitutive ("noninstrumental") dependence, put in one sentence: "[T]he dependence of perceptual *variations* on variations in output is noninstrumental when inputs are constant" (p. 363; Hurley's emphasis). Thus, the core idea of this kind of dependence is that perception differs with changes in output despite constant input. This seems to be a good description of the situation in the various experiments: the input stays the same across the conditions whereas perception alters with changes in output. 'Output' is to be taken in a general sense. According to Hurley, it spans from efferent neural impulses to motor signals and environmental effects. I take it that the subpersonal mechanism proposed by Proffitt et al., namely the motor simulation that underlies the effects can be regarded as 'output' in Hurley's sense. Therefore, it seems plausible to regard these cases as indicating a constitutive dependence of spatial perception on intended actions in terms of Hurley's notion.

105 Of course, strictly speaking, there is some movement involved in these experiments: reaching, throwing, or walking are all movements. However, these movements are not a means to change the retinal pattern to make possible certain perceptual contents that are not available without the movement, as the notion of instrumental dependence implicates.

However, I want to present a consideration that casts doubt on this classification. Hurley briefly mentions the possibility of an "internal instrumental dependence", which I did not mention hitherto since it is difficult to understand without a particular example at hand. I want to argue that it might be possible to explain the results of the Proffitt studies by drawing on this notion. This would indicate that the cases in fact merely show an instrumental dependence and not a constitutive one.

Unfortunately, Hurley does not say much about the internal instrumental dependence. She writes, "Intuitively, instrumental dependence might operate wholly internally: action might internally affect a separate perceptual system" (Hurley 1998, p. 381). She gives neither any further explication of it nor an example. The following is my reconstruction of the idea. The fundamental idea of the instrumental dependence is that a change in perception is brought about by a change in input that is caused by a change in output. Put differently, a certain output affects input, which in turn causes a perceptual change. Now, the idea of an internal instrumental dependence seems to be that this process happens entirely within the subject's boundaries as it were. Thus, motor signals affect sensory signals, which causes a change in perception. I take it that this is the core of Hurley's internal instrumental dependence.

For this notion to be intelligible, it is necessary to refine the notions of input and output. Hitherto, I analysed input primarily as the pattern of retinal stimulation. However, Hurley's account allows for a broader conception of input. The chain of inputs comprises everything between "environmental causes of sensory events through to afferent neural impulses" (p. 343; cf. p. 381). Thus, her notion of input comprises elements that lay beyond (environmental causes), on (retinal pattern) and within (afferent impulses[106]) the borders of the subject as it were. Analogously, the notion of output does not

106 The term 'afferent' is used as a predicate for any connections or information that proceed from the periphery toward the central nervous system. The term for connections/information that proceed in the opposite direction is 'efferent', which comprise e.g. motor signals; cf. Karnath and Thier (2012), p. 798. The dichotomy seems not to be exhaustive, i.e. there are neurons that code both sensory and motor related information; cf. Hurley (1998), p. 382, n. 37.

only refer to overt actions but also includes the neural motor signals that give rise to bodily movements. Given these broader notions of input and output, which are part of Hurley's account, the internal instrumental dependence between action and perception seems possible in the way formulated above. A motor signal affects a sensory signal, which causes a change in perception.

Is it possible to analyse the Proffitt studies in terms of this internal instrumental dependence? As explicated above, central to Proffitt's explanation is the notion of a motor simulation. This seems to be a good candidate for an output signal in Hurley's sense. Given this, the effects that actions have on perception in their experiments might be explained by the effect that motor simulations have on sensory signals. This would mean that the changes in perception are due to changes in sensory states that are brought about by motor simulations. Thus, the changes in perception would not take place while input is constant. Since this essential condition on a constitutive dependence of perception on action is not fulfilled on this interpretation, the Proffitt studies would not present cases of a constitutive dependence but of an instrumental dependence.

However, let me note that Hurley's notion of internal instrumental dependence is generally dubious. In the way I explicated the idea, it arguably would be applicable to the case of efference copies affecting perception. However, Hurley clearly takes this case as evidence for the non-instrumental dependence as her reference to the case of paralyzed eye shows (cf. section 2.2.2 above). In effect, the notion of internal instrumental dependence blurs the distinction between instrumental and non-instrumental dependence.

Finally assessing the question whether the Proffitt studies reveal an instrumental or a constitutive dependence would require not only further empirical work but also a more thorough analysis of the idea of internal instrumental dependence. The question that has to be answered empirically would be whether the motor simulations have an effect on perception via changing sensory input signals, or whether perception changes due to motor simulations while sensory input signals are constant. Only the latter case would exemplify a constitutive dependence, since only in this case perception varies with output while input is constant.

I cannot decide this question conclusively. Therefore, the possibility that the Proffitt studies indeed speak in favour of a constitutive dependence of perception on action is still open. In the final section on the Proffitt studies, let me consider the significance of this possibility.

3.1.2.3 The significance of the results of Proffitt studies

Suppose that the observed effects in the Proffitt studies indeed were examples for a constitutive dependence in terms of Hurley's framework. Thus, they would show that some cases of spatial perception constitutively depend on the subject's actions. However, what they would not thereby show is that the *capability* of spatial perception constitutively depends on action.

The reason for this has to do with the way Hurley sets up her framework. As already stated, her distinction of instrumental and constitutive dependence is put in terms of a distinction that applies to particular cases of perception and action (Hurley 1998, pp. 341–4). Given her basic aim, this is all she needs to make her point. She argues against the general application of the traditional 'Input-Output Picture' that states that perception is constituted by input alone while action is constituted by output alone and that there is a linear and one-way causal flow from input to output (see Ch. 2.2.2 above). To this end, she only needs certain cases that are not explicable based on it. In this context, she distinguishes between instrumental dependence (which is consistent with the Input-Output Picture) and constitutive dependence (which is not) and explicates certain cases that instantiate constitutive dependence of perception on action (pp. 344–62).

I think that the Proffitt studies (if they do show a constitutive dependence) have their theoretical place in such a project. They can show that some spatial perceptions are influenced by action. However, what they cannot show is a more basic and general relation between action and perception. They cannot help to show that the *capability* for spatial perception is dependent on action. It might well be that there are some particular cases of spatial perception that constitutively depend on action, but this would be consistent with the fact that the capability of spatial perception is not constitutively dependent on action. Furthermore, even if it turned out that action has an effect on all particular spatial perceptions, this would not show that action is a constitutive condition on the capability for spatial perception since it would not show that it is necessary for spatial perception. It would only indicate that action influences some of the spatial magnitudes of perceptual space but this would not mean that action is essential for spatial perception.

In sum, based on the Proffitt studies, one can only argue that action can influence, affect or modify spatial perception in certain circumstances, i.e. in a way that exemplifies Hurley's constitutive dependence. However, this does not mean that the capability for spatial perception is dependent on action. Since it is the latter claim that I am trying to explicate, the results that form the basic evidence in favour of the action-specific approach to perception certainly present intriguing effects, but are not, contrary to what one might have initially expected, relevant here.

3.1.3 Inverted vision

The second kind of empirical evidence I want to discuss are studies on inverted vision. Studies of this kind have been discussed for more than hundred years and are still controversial. The first experiments on inverted vision were conducted more than a century ago by George Stratton (1897) when he was a student of Wilhelm Wundt in Leipzig (cf. Wade 2000, p. 254). The most recent ones are reported by Kaoru Sekiyama et al. (2012), and Jan Degenaar (2014).[107] Although they vary with respect to the equipment that is used and in the ways vision gets inverted (left-right, up-down, or both ways), the observed phenomena are roughly the same.

Two observations are most important. The first is that after putting on the inverting device, participants quickly manage to interact with their environment based on what they see. Roughly after three days of exposure to inverted vision, subjects are able to interact with perceived objects almost normally in

107 Other studies are reported in Kohler (1951; 1964) (cf. Kohler's introduction for more references to experimental work in the first half of the 20th century); Taylor (1962); Dolezal (1982); Linden et al. (1999).

terms of speed and accuracy of their movements. Thus, the first observation is that sensorimotor skills, i.e. the ability to use visual information for action control and guidance is re-instantiated in quite a short amount of time. It is important to note that at the time during which sensorimotor behaviour is as good as before, the visual field still is inverted. Thus, subjects can guide their actions properly despite a fundamental distortion of their visual experience.[108]

The second observation is that in most cases phenomenal experience is re-inverted after sufficient long exposure to inverted vision.[109] However, the reports are not as clear on this point as with respect to the re-instantiation of sensorimotor skills. The reports not only differ about the timespan that is sufficient for the phenomenal experience to change (there seem to be great interpersonal differences here in this respect)[110]. They also differ with regard to the exact description of this change[111] and the extent of the alteration: in some cases, only parts of the perceived environment are re-inverted, while most investigators report that finally the entire visual space re-inverts and is thus almost the same as in visual perception without inversion.[112]

Overall, the experimental findings suggest that subjects not only learn to use highly distorted visual information for action guidance but also that, given a sufficiently long exposure to inverted vision, perceptual space gets re-inverted.

Let me remark that the second observation with regard to inverted vision studies provides *prima facie* evidence for one central claim about space perception that is endorsed by accounts that take action as a constitutive condition

108 For example, Linden et al. (1999) report that subjects "are able to interact correctly with the surrounding world after a few days" (p. 480) despite no return of upright vision. All other studies report the same observation for at least some periods during the experiment.

109 One notable exception here is Linden et al. (1999), where no phenomenal re-inversion is observed. However, this may be due to the relatively short durance of the experiment, as is suggested by Sekiyama et al. (2012).

110 This is specifically obvious in the study of Sekiyama et al. (2012), where one of the four participants showed experiential effects only very late (after a month) while the others showed effects after roughly two and a half weeks.

111 Dolezal (1982) e.g. reports that the resulting perception was unfamiliar or weird and not like before vision was inverted. Degenaar (2014), p. 387 speaks of a "Gestalt switch".

112 Cf. e.g. Degenaar (2014) and Stratton (1897) for these observations.

on spatial perception. This is the claim that the pattern of proximal stimulation alone is not sufficient to account for the structure of perceptual space. If it were, phenomenal space would not re-invert after a sufficiently long time span of exposure to inverting devices. The retinal image stays inverted while perceptual space gets re-inverted. Thus, it is clear, that proximal stimulation alone does not exhaustively determine the spatial structure of visual perception. There has to be some additional element invoked to explain the re-inversion.

In the following, I want to consider in what way these cases speak in favour of the constitutive view. In particular, the question is how these views can explain the first observation that perception-action links are established without a concomitant change in visual spatial experience. I will first argue that it cannot be explained within the sensorimotor framework proposed by Noë. In fact, it presents an objection to that approach. Second, I will argue that these cases fit approaches that take the capability to act as constitutive. In Ch. 2.3, I primarily discussed two such accounts, from Evans and Scheler. I will show how the second observation can be explained within each of their approaches.

3.1.3.1 A challenge to the sensorimotor view

The results of the inverted vision studies present a problem for the sensorimotor approach, as expressed in Noë (2004) (cf. Ch. 2.2.3 above), for the following reasons. In case the basic claim of this approach was right – namely that perceptual space is constituted by knowledge about how sensory patterns shift with movement – one would expect that, as soon as the relevant sensorimotor knowledge changes, perceptual space changes, too. However, as stated above, all studies show that roughly after three days, subjects are able to act fluently on what they see while their visual field is still being inverted. This indicates that new sensorimotor knowledge is developed which is the basis of the acts. However, there is no change in perceptual space. Therefore, the findings speak against the sensorimotor approach to perceptual space (cf. Klein 2007).

To further illuminate this point, consider the following statement from Noë (2004) that is directed against views such as Evans's: "The experience of the flicker as on the visual egocentric right consists not in our disposition to move in certain ways, but in our possession of a kind of practical knowledge of how movement would bring the thing into view" (p. 89). It seems that subjects in the studies have this kind of knowledge. Presumably, they can orient toward seen objects. For example, they know that they have to turn their head to the right in order to face an object they see to their left. This indicates that their sensorimotor knowledge has changed without a concomitant change in perceptual space. Thus, the cases speak against the sensorimotor approach.

3.1.3.2 Inverted vision studies as evidence for capability approaches

Let me consider how the results of the inverted vision cases can be used to argue for the capability approaches. I will consider separately how they fit the accounts of Evans and Scheler. The accounts face the same challenge, as did the sensorimotor approach, consisting in the first observation stated above. The problem is that perceptual space is not re-inverted as soon as subjects are able to interact with what they see.

As discussed in Ch. 2.3.3, on Evans's approach, dispositions to act that are directly linked to perceptual input partly constitute perceptual space. I think there are two ways in which the inverted vision studies can be explained within this kind of account.

The first has been proposed by George Pitcher (1971). He argues that subjects are disposed to *a plethora of different* behavioural responses to one sensory state. The fact that one can interact with a given object without perceiving it at its actual location is explained by the claim that only few dispositions are linked to the perceptual information that result in successful behaviour whereas the majority of dispositions would result in unsuccessful behaviour when actualized. Therefore, since most of the movements one is disposed to are directed at the wrong location, the object is accordingly perceived at the wrong location (pp. 159–60). Effectively, this reply to the objection consists in the claim that perceived location is determined by the endpoints of the majority of behavioural dispositions.

This reply might be possible but, on the face of it, it sounds *ad hoc* as it stands. It relies on the idea that one can somehow count the various dispositions to act that one has at a time. For this to be possible, it must be explained how the behavioural dispositions in question are individuated (in terms of e.g. grasping vs. kicking an object, or rather in terms of e.g. grasping *this* way vs. *that* way?). Pitcher does not offer criteria for individuating dispositions. This does certainly not rule out the proposal, but it highlights the need for substantial extension of it to be convincing.

Furthermore, it must be explained why a solution in terms of quantity is superior to other possibilities. One such alternative proposal would be to distinguish between different kinds of dispositions and argue that perceptual space changes with a change in a certain kind of dispositions. This is the second reply to the challenge that I want to discuss that seems open for proponents of Evans's approach.

On this proposal, it may be that what subjects perceive in the relevant stage of the experiment is linked to certain dispositions to move although when they actually react or interact with the world they execute different movements than those the perceptual information disposes them to. The situation might be this: visual information is linked to certain dispositions to move that, when actualized, would not be successful. (One sees an object to the right because one is disposed to extend one's arm to the right in case one would want to grasp it, however, the action would not be successful since the object really is to the left of oneself.) On the other hand, when one is actually interacting with the objects around oneself one relies on already established unconscious sensorimotor links. That is, the actual executed processes that lead to one's behaviour are attuned to the new situation when the dispositions one has to act on what one sees still follow the older schema.[113]

113 In refining the relevant idea of disposition this proposal is akin to Peacocke's (1983) strategy of refining Evans's position. However, note that he ends up with a different notion of disposition that includes the movement of the organism to different places.

On this proposal, two kinds of disposition need to be distinguished to explain the cases. In general, a disposition describes how something would 'react' to the presence of certain circumstances. 'React' can mean various processes. "The frog is disposed to extent its tongue thus and so when he spots a fly in a certain way", or "Homer Simpson is disposed to grasp a can of Duff Beer whenever he sees one", are both viable descriptions of dispositional properties.[114] For explaining the case of inverted vision, one might claim the following. One kind of practical dispositions that concerns the habitual, subconscious dealing with the world quickly attunes to the new situation and lets subjects interact with objects in normal ways. However, dispositions of another kind, namely those that concern consciously controlled, planned behaviour are slower in adapting.[115] Whereas the former dispositions to act are linked to our largely unconscious and automatic interactions with the world, the latter are associated with higher-level notions of acting.[116] Finally, on this proposal, the dispositions of the second kind are the ones that determine the spatial structure of perceptual space.

I am not sure whether this proposal is already beyond Evans's account, since he only speaks about behavioural output being linked to perceptual input without suggesting any difference among the dispositions one has. However, I think this extension of his account would be more reasonable as Pitcher's proposal in terms of the quantity of dispositions.

Finally, let me turn toward a reply to the challenge that is based on Scheler's approach. Remember that, whereas on Evans's account the factual presence of dispositions that are directly linked to perceptual information is crucial, on Scheler's account it is some basic grasp of what movements are possible that constitutes perceptual space. The difference is thus between a

114 A third sense might be exemplified by cases such as "Water is disposed to freeze when the surrounding temperature is at or below 0° Celsius." These are not relevant here.

115 The process of adaptation might lead to strange spatial perceptions like seeing an object at two locations at a time as reported by Kohler and Taylor (cf. Degenaar 2014, p. 392 for references). However, this observation is not made by all investigators.

116 The distinction might be akin to the one that Place (1996), p. 116–7 draws between what he calls "'physical' dispositions" and "biological dispositions".

mere factual possibility and a possibility of which the subject is somehow aware.

One might argue that the critical stage we are trying to explain reflects the fact that the subject is not aware of the practical capabilities that are open to him in response to what he perceives in the right way. Thus, according to this proposal, the fact that the subject is able to interact with objects in his environment successfully is not sufficient to ground perceptual space since what in fact is constitutive of spatial perception is something more than the mere possibility of interacting successfully: it is the presence of the grasp of this possibility. Hence, the perceptual re-inversion is dependent on this basic grasp of what one can do; something that comes later than actual successful movements and might come in degrees. This would not only provide an explanation of the relevant observation on which the challenge rests but also of the sometimes reported partial re-inversion of perceptual space and the even fewer reported cases of seeing things at two different locations at the same time.

In this section, I have argued that both, Evans's and Scheler's account can accommodate the problematic observation made within studies on inverted vision, namely that there is a stage in the process of adaptation to the new situation where subjects are able to interact with their environment while a change in perceptual space occurs only afterwards. It seems that Scheler's response is more straightforward since on this proposal, it is possible to argue that the grasp of the new action possibilities with respect to what one sees takes some time to get updated. In contrast, Evans's approach would have to be slightly extended to accommodate the findings, either by dividing the behavioural dispositions quantitatively or a qualitatively. I sketched both answers and argued that the second is more promising.

Finally, let me emphasize again that the studies on inverted vision in general present an important piece of evidence for constitutive views. They show that the way we perceive the world spatially is not only dependent on the way the light hits the retina but also that it is flexible to a certain extent. One straightforward explanation is that the spatial orientation of perceptual space varies with the dispositions to act in one of the manners outlined above.

However, finally, I briefly want to consider an alternative explanation that relies on the integration of multisensory signals and accounts for the perceptual change without referring to action.

3.1.3.3 An alternative proposal

There is at least one other reasonable possibility to explain the observations about inverted vision cases. One might hold that what essentially happens in inverted vision cases is the destruction of a 'harmony' that is so basic to our normal living that we hardly think about it at all. What I mean is that the perceptions we acquire on the basis of our various sensory systems are each structured in their distinctive spatial scheme and yet in usual cases our experience is of one unified space. We hear things at the same direction at which we see them and haptically sense them, although visual space is indexed to the eyes, auditory space to the ears and haptic space to the relevant effector. In cases of inverted vision, visual space is disharmonized from the other perceptual spaces. One does not see things in the direction where one hears them anymore; one does not touch them where one sees them, and so on.

However, one still knows that what one perceives through one's various senses is just one identical world. One cannot solve the tension between the conflicting spatial structures by assuming that what one sees is something completely different from what one otherwise perceives. Therefore, there has to be some way of re-harmonizing one's various perceptual spaces. One cannot live constantly in a self-contradictorily presented world. The pressure for solving this tension is supposedly high enough to be effective even if one is capable of acting on such an incoherently presented world.

What would be the most efficient way to solve the conflict between the various perceptual spaces? Surely, to change the space that is not in harmony with those that are with respect to each other. Since auditory, haptic and proprioceptive spaces are still consistent, the obvious solution is to change visual space. This means that the spatial scheme of the other perceptual modalities somehow serves as the paradigm against which the incoming information from the eyes is measured. In the course of the experiment, the participants learn to form visual representations that are consistent with their representations of the outer world that are based on non-visual information. Of course, this learning process is not to be regarded as something that is consciously learned in a planned manner by the subjects. It is a rather automatic process that leads to an increasingly consistent representation of the world as presented through the senses (cf. Held 2009; Rochat 2004, Ch. 2).

On this proposal, the re-inverted spatial structure of visual perceptual space is constituted by the other already present and harmonized perceptual spaces and some pressure on us to avoid contradicting spatial representations and to form a consistent representation of the world across different modalities. However, Held argues that active movements of the subject facilitate the remapping. It is important to note, though, that on this proposal action plays only an enabling role to make possible the successful remapping of the various space to generate a harmonized overall spatial representation again. Action is not taken as a constitutive condition on spatial perception here, but merely as a means to align the various sensational spaces. The changes in the orientation of distorted perceptual space are explained by the remapping processes. Hence, this proposal explains the inverted vision cases without taking action as constitutive on spatial perception, it seems.

However, there are at least to possible answers to this proposal, which proponents of a constitutive view can provide. First, one crucial maxim is that we must not confuse constitutive and developmental issues. The above outlined proposal for an explanation of the re-inversion presents a possible developmental story of how visual perceptual space gets re-inverted by a remapping of the visual space. From such a story, we cannot straightforwardly infer anything about the constitutive conditions on visual spatial representation. Furthermore, this point also indicates that the fact that action facilitates the remapping of the spaces cannot be used as evidence for the constitutive claim.

Furthermore, even if appealing to perceptual spaces of other modalities can explain the re-inversion of visual space, this does not rule out that spatial perception *per se* is constituted in part by connections to action. The proposal does not explain the constitutive conditions of spatial perception as such but relies on already established perceptual spaces. Hence, a successful explanation along these lines is certainly no knockdown argument against the constitutive claim.

3.1.3.4 Summary

In sum, I think that the results of inverted vision studies present an important body of evidence for the claim that action is a constitutive condition on spatial perception. I argued that the two most important observations about these cases are that the subjects quickly learn to use highly distorted visual information for action guidance and that, given a sufficiently long exposure to inverted vision, perceptual space gets re-inverted. The second observation provides evidence for the fundamental conviction of approaches that take action as constitutive in that it shows that spatial perception is not determined by spatial properties of the retinal image alone. However, the first observation is problematic since it highlights a temporal gap between the ability of visually-guided actions and the change in the orientation of the visual field. I argued that this is especially a problem for sensorimotor approaches, since these imply that spatial perception would change as soon as the (new) sensorimotor knowledge is present. In contrast, capability approaches, and in particular Schelerian approaches, fare better in accommodating the results. Finally, I presented an alternative explanation of the re-instantiation of the visual perceptual field that draws on remapping processes. However, this proposal presents only a developmental story about how the changes in perceptual space might come about. It does not commit to any constitutive claims about spatial perception in general.

Inverted vision studies can be used only to argue that action is a condition on the distinctive spatial organization of perceptions. They are not useful to argue that action is a constitutive condition on the ability of spatial perception. This is because they reveal how action has an effect on already established (if distorted) spatial perception. In the next section, I will present empirical evidence from studies on sensory substitution devices that speaks in favour of the claim that action is a constitutive condition on the *capability* for spatial perception (cf. chart in Ch. 3.1.1). Thus, in these cases we will see a transition from non-spatial to spatial perception as it were.

3.1.4 Sensory substitution devices

The last kind of empirical evidence for the claim that action is a constitutive condition on spatial perception I want to consider stems from considerations about the functioning of sensory substitution devices (SSDs). The use of these devices *prima facie* presents a highly relevant case since it provides an opportunity to directly compare differences between perceptual states have spatial significance with those do not, while both states are caused in the same way. Defenders of the view that action is constitutive of spatial perception hold that in one or another way the capability to act on the relevant perception is constitutive of the spatial significance.[117]

In general, SSDs are a means to compensate for the loss of a sensory modality, i.e. they help subjects who have lost their vision, audition, touch or vestibular sense to compensate for the loss by (in most cases) using another sensory modality. They come in different varieties.[118] The most extensive studied devices are those used to compensate for loss of vision. There are two kinds of SSDs that are predominantly discussed in the field. Both are intended to help the blind 'see' in a way. The first kind of device works by translating signals gathered by a video camera into tactile stimulation, on either the back, abdomen, or tongue of the subject.[119] The second kind works by translating

117 Sensory substitution is used as evidence for such a view by e.g. O'Regan and Noë (2001), Hurley (1998), Grush (2007), and Briscoe (forthcoming).

118 For an overview, see Bach-y-Rita and Kercel (2003). Contrary to what almost every characterisation of SSDs holds, not in any case a literal 'substitution' of one modality for another takes place. For example, Paul Bach-y-Rita, arguably the most important researcher in the field, also lists a device that compensates for the loss of touch in the fingers with tactual sensations on the forehead where the relevant tactual information is transmitted by special gloves; cf. Bach-y-Rita and Kercel (2003), p. 541.

119 For every kind of device, there are many different particular devices that differ in the way the signals are translated. For overviews of work with tactile substitution of vision, see e.g. Bach-y-Rita (1983); Bach-y-Rita et al. (2003); Bach-y-Rita (2004); Renier and Volder (2005).

video-captured signals into sounds.[120] Particular devices vary in the way they translate the information.[121] However, they all have in common that patterns of spatial (and sometimes other kinds of) information that are received via the camera are translated into distinctive patterns of auditory or tactile sensation. The spatial information that is thereby made available to the subjects allows them to discriminate different objects as well as their location in the surrounding space.

It has to be noted that the term 'sensory *substitution*' is not exactly appropriate. No one of the researchers actually claims that the usage of an SSD leads to the (re-)instantiation of e.g. vision that is indistinguishable from vision of healthy subjects (Matthen et al. 2013). There are important features of normal vision that are lacking in sensory substitution. For example, the overall phenomenology seems to be different from the one of vision as far as one can tell. In addition, it is observed that perception grounded in sensory substitution lacks affective qualities (Dokic 2013). This has led to a large discussion about the question of which sensory modality the resulting sensations are of. The options that have been discussed in the literature range from taking the resulting perception as an instance of the substituted modality, or of the substituting modality, or else to take it as a whole new kind of modality that is instantiated (Farina 2013). There seems to be some agreement in the relevant scientific community on thinking about what is going on in sensory substitution in terms of an augmentation of the substituting modality instead of a (re-)instantiation of the lost one (Matthen et al. 2013).[122]

However, as important as this question might be for some theoretical purposes like an account of the individuation of sense modalities, it is not decisive

120 For overviews cf. Bach-y-Rita and Kercel (2003), p. 542; Heil (1987), pp. 235–9; Renier and Volder (2005); Meijer (1992); Capelle et al. (1998).

121 To give at least one example: In the case of "The vOICe" (the middle part means "Oh, I see!"), a device developed by Peter Meijer, elevation of stimuli varies with pitch, brightness with loudness and direction of the source of stimuli with stereo sound. For more details, sample sounds and videos cf. http://www.seeingwithsound.com/, checked on 18/08/2014.

122 Another proposal in this debate with which I am sympathetic is made by Martin and Le Corre (2015), who argue that the resulting experience is best described as purely spatial.

for my purposes. Whatever the right modality ascription finally is, the most important observation is that people using SSDs not only acquire the capability to purposefully interact with thereby perceived objects but they also report that they do perceive objects in the space around them instead of perceiving just the patterns of tactile stimulation or sound after they got used to the devices.[123]

I take it that there are two critical questions here that have to be settled in order to provide the possibility of taking sensory substitution as evidence for the constitutive dependence of spatial perception on action. The first is whether there is genuine spatial perception involved in these cases at all. The second is in what way these cases can then be used to explain the constitutive dependence of spatial perception on action. I will address these questions in turn.

3.1.4.1 Spatial perception or judgment?

The first question sounds rather preliminary, although it is as important as the second one. Without showing that there is genuine spatial perception in these cases, the whole case would be a non-starter. The issue here is whether the spatial significance that is present in sensory substitution really is part of the perception or whether it is a cognitive accomplishment. One could argue that subjects using SSDs do not literally perceive things as located in space around them but that they simply infer[124] where an object is on a particular occasion from the experiences they have in the substituting modality and learned associations between these and locations of objects around them (Macpherson 2013). If this were what is going on, subjects would not perceive things as being located in space. They would just perceive patterns of tactile or auditory stimulation and infer from these the location of objects. Since there would not be any *perception of a distally located object* in the first place, these cases could

123 For an exceptionally fascinating report of a long-term experience with an auditory SSD that points in favour of taking the resulting perception as genuine vision, see Ward and Meijer (2010).

124 The relevant notion of inference here refers to rational transitions between mental states on the *personal* level. (Thus, not in the sense it is used in theories of perception inspired by Helmholtz that centre on sub-personal 'inferential' processes.)

not be used as evidence for the constitutive claim. The phenomenon, whose constitution shall be explained, would not be present.

I think this issue has not received enough attention by those who use SSDs to argue for the constitutive claim. For example, O'Regan and Noë (2001), or Grush (2007) simply take for granted that SSD-users enjoy spatial perception: "It is clear that while such devices obviously cannot provide visual experience, they nevertheless provide users with the clear impression of things being 'out in front of them'" (O'Regan and Noë 2001, p. 956).[125] Since a plausible alternative is available, it is necessary to first argue for the presence of genuine perceptual space (including distal attribution) in these cases. As will become apparent, this is far from being a trivial task since evidence in favour of a proper spatial perceptual character of the experience in sensory substitution is not easy to find. I will draw on four different sources, which are subjective reports, a special observed behavioural effect (the looming effect), research on visual illusions and two studies that investigate distal attribution in sensory substitution. I will finally claim that at least under certain circumstances there is genuine perceptual space present in these cases although the evidence for that claim is by far not as unambiguous as is widely claimed.

Subjective reports

The hypothesis that merely a cognitive or inferential process leads to the 'impression' that there is spatial perception present in sensory substitution is supported by reflecting on what is going on when someone wears the device for the first time. Suppose the subject is informed about the functioning of an auditory substitution device for vision, and so knows in what ways properties of the sound pattern such as pitch, volume, apparent horizontal direction, and other qualities covary with properties of an object or the scene in her surrounding space. Based on this knowledge the subject is supposedly able to infer spatial properties and locations of things around her and even to identify objects and people based on focussing on the sound pattern and using her background knowledge about the relevant covariations to infer the distal

125 Note that the term 'impression' here is ambiguous.

source of the stimulation.[126] However, this does not show that there is genuine perception of the object being located in distal space. The subject might just *know* it is 'there' based on her reasoning, without *perceiving* it to be 'there'.

Fiona Macpherson (2013) has pointed out that the documented reports of subjects using SSDs are generally compatible with this scenario. One prominent example here is the report of a congenitally blind subject given in Guarniero (1974). He writes that after a three-week training with a visual-to-tactile substitution device, "objects had come to have a top and bottom; a right side and a left; but no depth – they existed in an ordered two-dimensional space, the precise location of which has not yet been determined" (p. 104). Despite the fact that he was not aware of the stimulations on his back as he was at the beginning of his training (ibid.), he did not yet perceive objects as being located "out there" (p. 101). Therefore, he had to rely on some cognitive strategy to infer from his perception that its source would lie at some particular distance in the vicinity.

What can be said in response to argue that SSD-users actually perceive objects in space and do not merely know somehow that they are there? For one thing, a great number of reports of people using SSDs state that they literally perceive things as located somewhere in the vicinity. Bach-y-Rita (1984) reports that the "change from the perception of a tactile stimulation to a spatial three-dimensional perception occurred in most TVSS [tactile-visual sensory substitution, F.P.] subjects after 5–10 h of training" (p. 156). First-hand reports also reveal that there is more than mere proximal stimulation going on in these cases. Even Guarniero, who claims to have no spatial perception, reports that the proximal tactile stimulation became more and more transparent and that he perceived objects with distinctive shapes.[127]

---

126 Levy-Tzedek et al. (2012) show that accurate reaching movements with an SSD are possible after short practice with the device; cf. also Brown et al. (2011) and Auvray et al. (2007).

127 "Transparency" is notoriously a multi-faceted concept. In the present context, it means the fact that the perception in the substituting modality becomes more and more of a background phenomenon (in a phenomenological sense) and vanishes eventually completely while at the same time functioning as a 'medium' that reveals the distal sources of this perception to the subject.

Thus, Guarniero's description is quite interesting since it does not seem to fit the often-made dichotomy between proximal stimulation and distal perception. On the one hand, he reports that the tactile sensation on his back has become transparent for him during the training, i.e. he does not merely perceive patterns of pressure on his back but "objects" with a distinctive shape. On the other hand, these objects are not perceived as being at a distinctive location in the space around him. Although this clearly suggests that he had no proper spatial perception of objects, it does suggest that there is at least some form of perception involved, thus supporting at least the claim that sensory substitution is not exclusively a matter of cognitive inferences from proximal sensations to distal sources. He does perceive objects even if not located at a distinctive location, hence at least some perception, or better, some fundamental shift in his perception from tactile sensations to object perception is involved in his case, too.

Guarniero used the device only for three weeks. Maybe in his case this simply was not enough to establish proper 3-D-perception? Support for an affirmative answer to this question comes from the report of the blind subject PF in Ward and Meijer (2010). PF is one of the very few subjects that has used an SSD over a long period. She has used an SSD called 'The vOICe'[128] continuously and immersively since roughly the year 2000. In accordance with Guarniero's report, she claims to have seen no depth in the first few months of using the device. However, after about one year of steady use she suddenly began to experience depth:

> [And] I walked slowly through my house looking into the rooms and it was like incredible. I could see into the room. That flat drawing now has depth to it. I can sense it. I can't tell you what is across the other end of the room but I can tell you that it is across from me. (Ward and Meijer 2010, p. 496)

One is tempted to think that the "flat drawing" of which PF reports here refers to the same kind of experience that Guarniero (1974) described as an "ordered two-dimensional space" (p. 104). Since PF used her device for much

128 Cf. p. 169, note 121 above.

longer, one might say that the lack of real depth perception in Guarniero's case is just due to the short time he used the device.

However, issues are much more complicated than they seem *prima facie*. There are many dissimilarities between the two cases. They used different kinds of devices (tactile and auditory) and used different training methods. In addition, one has to keep in mind that there are individual differences in the experiences subjects get through SSDs (e.g. Bach-Y-Rita speaks of "most" of his subjects gaining depth perception). The crucial point certainly is that Guarniero was blind from birth whereas PF only got blind at the age of 21 (Guarniero 1974, p. 101; Ward and Meijer 2010, p. 494). Although she had been blind for more than 20 years before starting to use the device, it is plausible to suppose that her former visual experiences facilitated the establishment of depth perception (p. 497). Therefore, one might suspect that her experience is best analysed as an imagination and not a perception of her surroundings.

In sum, subjective reports seem to be an unreliable basis to some extent for deciding this matter. Most of the reports are consistent with an inferential picture; one prominent example explicitly states that no proper spatial perception occurs; another states that it does, while this might be explained by visual imagery based on former visual experiences. Overall, subjective reports thus establish only weak evidence.

Are there other means to decide whether distal attribution occurs in sensory substitution cases aside from subjective reports? I want to consider three other sources of evidence: the first is the repeatedly cited 'looming effect', essentially a behavioural response to a sudden change in proximal stimulation; the second is the putative presence of visual illusions in sensory substitution; and the third comprises two studies that try to investigate whether distal attribution is present in sensory substitution. I will discuss these in turn.

The looming effect

One rather anecdotal evidence for the perceptual character of sensory substitution is the instantaneous reaction of subjects using an SSD to approaching objects. John Heil (1987) reports that infants using a vision-to-audition SSD, the 'Sonic Guide', for the first time dodge approaching objects they perceive with the device (p. 239). It is not only somewhat surprising that they react in that way "without any practice or training at all" (ibid.), it does also militate for the claim that they perceive an approaching object in 3-D space, including depth. A looming object that would be perceived merely in an ordered two-dimensional space would just change its size, become larger, and hence would unlikely lead to a dodging reaction.

Similarly, Bach-Y-Rita (1984) reports that as an instructor one day moved the zoom control of one subject's camera without him knowing, the subject showed an immediate defence reaction. The subject obviously perceived an approaching object based on the sudden enlargement of the proximal stimulation caused by the zooming. Moreover, the subject in this case received a tactile stimulation on his back whereas his reaction clearly was to an object coming from ahead (he raised his arms and stepped back). This seems to show that the subject had a perception of an approaching object, coming from ahead. It is clearly no reaction towards the tactile sensations on his back. This strongly suggests that the perceived stimuli were transparent for the subject and allowed him to locate (moving) objects in the space around him.

However, as Macpherson (2013) points out, one might still hold that there is a fast inference drawn based on the changing pattern of proximal stimulation and that the reaction thus does not show that there is genuine perception of an approaching object. Support for this claim might come from the facts that the subject in Bach-Y-Rita's report had a similar experience a few days before, where an object fell on him, and where he did not show the defence reaction (Bach-y-Rita 1984, p. 156), and that he was overall a "well-trained subject" (O'Regan and Noë 2001, p. 956). In light of this, one might argue that he simply recognised the sudden change in proximal stimulation to be the same as before, thereby knowing what would happen based on his former experience and acting accordingly. From this point of view, it is not necessary to assume genuine distal perception to explain the case. One might insist that the immediacy of his reaction speaks against any inferences. However, time is a difficult measure to decide here, I think. It strikes me as impossible to judge whether this reaction was based on perception or inferences only by regarding time. Furthermore, even if the subject would report that he did not rely on any inferences, it is still possible that there was either some unconscious, fast, automatic inference going on here or the reaction simply is to be regarded as a learned association, viz. a reflex triggered by the pattern of stimulation. Overall, pace O'Regan and Noë, and Bach-Y-Rita, it does not follow from this anecdote that the subject really perceived some approaching object.

Perceptual illusions

Until now, I presented evidence that is at best compatible with the claim that there is genuine spatial perception in sensory substitution. We are still lacking some convincing case in favour of it. One appealing way to argue for it would be to find perceptual illusions that are present in sensory substitution. Macpherson (2013) claims that at least two perceptual illusions can be present for subjects using a tactile SSD: the Müller-Lyer-Illusion and the waterfall illusion. Unfortunately, as far as I know, there is no detailed report of an investigation of these matters to be found beyond anecdotal evidence given by Bach-Y-Rita that these illusions are present (Bach-y-Rita 2004, p. 86; Bach-y-Rita and Kercel 2002, p. 291; Bach-y-Rita 1984, p. 158; Bach-y-Rita 1972, p. 157). The presence of these illusions would indeed be good evidence for the genuine perceptual character of the resulting experience.

Perceptual illusions are generally present even if the subject has competing evidence. In the Müller-Lyer-Illusion, one perceives the two lines as of different length although one believes that they are of the same length. If this contrast between one's perceptions and one's beliefs was present in sensory substitution cases it would tell against the claim that what the subjects report is only based on sensing vibration on their back plus inferences. However, since there are no systematic investigations of the presence of these illusions, I do not want to lay too much weight on them here. Furthermore, although the presence of certain perceptual illusions might be good evidence, one has to be very careful in drawing any inferences from the presence or absence of illusions in general. For one thing, it is known that not all illusions are present in sensory substitution. Laurent Renier et al. (2005; 2006) have shown that the Ponzo Illusion and the vertical-horizontal illusion are not observed in every case of using a vision-to-audition SSD (PSVA). In particular, whereas only sighted but blindfolded subjects using an SSD perceive the illusion and to a lesser degree than sighted non-blindfolded subjects, early blind subjects do not at all. Another issue is that many visual illusions have haptic counterparts so that blind people can feel e.g. the Müller-Lyer-Illusion and others by touch (Heil 1987, p. 235; Casla et al. 1999; the latter failed to find the presence of the Ponzo illusion, however). Therefore, one cannot in any case infer from the presence of some illusion in tactile sensory substitution that subjects perceive anything different from tactile sensations. This means that from the presence of the illusions one cannot simply conclude that there is some distal attribution present. One might infer from these cases that there is genuine perception in sensory substitution but this does not mean by itself that there is also distal attribution present. What we need is evidence to the effect that subjects can perceive things in space around them, at some direction and distance from them. However, the presence of *these* illusions might be explained by the tactile perceptions the subjects have from the SSD. What would be needed for our purposes is the presence of an illusion in sensory substitution that involves 3-D spatial perception. This might be possible but as far as I know, there is no reported investigation to be found yet.

Experimental studies on distal attribution

Finally, I want to turn to two recent studies that investigate the question of whether there is distal attribution in sensory substitution (Auvray et al. 2005; Siegle and Warren 2010). The presence of distal attribution in these cases would mean that subjects using SSDs do not merely have auditory or tactile sensations, but perceive the sources of these sensations as objects in the vicinity. In that sense, the presence of distal attribution would be sufficient for the presence of genuine spatial perception of the vicinity, since it would indicate that what is perceived is attributed as properties to an object that is located at a certain distance and direction from the subject. However, in both studies, the notion of distal attribution is interpreted differently. Siegle and

Warren (2010) define distal attribution as "the perceptual experience of objects in external space" (p. 208). In contrast, Malika Auvray et al. (2005) analyse distal attribution as a gradual phenomenon that consists of three components: realization by the subject of a sensorimotor coupling, realization of an outer object as the cause of the sensations, and the presence of perceptual space.

Whatever the right notion of distal attribution may be, we have to get clear on what we need the studies to show to be of any value for our purposes. Remember that the issue here is about spatial perception. Therefore, we clearly need distal attribution in Siegle and Warren's sense, or in the sense of Auvray et al.'s third component. The mere presence of the other two components would be compatible with there being no genuine spatial perception (presence of merely the first component seems compatible with no distal attribution at all).[129]

The authors of both studies (to my mind rightly) think that the question of distal attribution in sensory substitution has not yet been investigated properly.[130] However, they give different reasons for that claim. Auvray and her colleagues highlight that in former experiments with SSDs, subjects already knew about the functioning of the devices and that there is a distal source of the sensations they gather via the devices. Siegle and Warren, in contrast, assume that distal attribution can be a matter of both perception or inference (cf. discussion above) and they see the great shortcoming in former studies in not distinguishing between these two options. The particular way in which both teams of authors tackle the issue of distal perception differs accordingly. The question that Auvray et al. are trying to answer is whether

129 The situation might be different if one wanted to use these examples to argue for the presence of spatial representation. It seems that the presence of spatial representation is considerably more certain than the presence of *perceptual* spatial representation in these cases. This is because the presence of spatial representation is given in case the subjects merely acquires some beliefs about the spatial properties of their environments, which might be based on learning what patterns of tactile or auditory sensation are caused by what objects around one, something that is compatible with there being no *perceptual representation* of space proper. However, since I take it that the case for the presence of genuine spatial perception in sensory substitution is not desperate, I stick to the argumentation in terms of perceptual space.

130 One exception is cited by both: Epstein et al. (1986).

subjects also attribute the cause of their sensations to some outer source in case they do not know that they are receiving signals from an SSD. In contrast, the goal of Siegle and Warren's study is to find out whether distal attribution is indeed a perceptual phenomenon as claimed in the literature.

Note that both questions are unrelated to a certain extent. Showing that distal attribution can occur in the absence of prior knowledge does not *per se* show whether distal attribution is a perceptual or an inferential accomplishment. Thus, regarding Auvray et al.'s study, it would be essential to look at the instantiation of the third component of their notion of distal attribution, i.e. the presence of perceptual space. Unfortunately, it is dubious whether they really could provide evidence for the presence of perceptual space given the methodology of their study. Subjects wore head-mounted video cameras without knowing it and manipulated some object either directly or via a lever. Their manipulations of the object were recorded by the camera and translated into sounds they heard with headphones. In effect, they were using an SSD without knowing it; they heard sounds that were caused by their own movements without being aware of this fact. After 20 minutes of using the device, subjects should judge the plausibility of seven distinct scenarios of what was the case in the situation. However, they were not asked about their experiences. Hence, in general, with this method one cannot distinguish whether the rating of a scenario that includes the notions of object and space is grounded in their perceptual experience or in some consideration or inferences about how the sounds were produced. Hence, it seems to me that their experimental means were inappropriate to study the presence of "complete distal attribution" (p. 507), including perceptual space, contrary to what the authors claim (p. 518). Even if subjects would have rated the actual situation they were in to be the most plausible one, it is not clear whether they would have done so because they enjoyed a spatial experience or because they reasoned that this was the most plausible explanation for how the sounds were produced.

Hence, the only thing that can be concluded from the results of this study is that subjects do, in the right circumstances, attribute the source of their sensations to objects in their environment. Auvray et al. found that this was only the case when subjects were able to put an obstacle between them and the perceived object, i.e. when they learned that the "information channel" could be interrupted. Doing this led to significantly heightened ratings of those scenarios that included the idea that there is an object before them that causes the sensations. However, as was already said, this result alone does not imply that subjects did literally perceive an object in space. Whether this might have been the case was the goal to investigate by Siegle and Warren's study, to which I now turn to.

In their study, subjects knew about the functioning of the device. They used a tactile SSD that converted signals gathered by a small video camera mounted on the subject's finger into tactile stimulations on the subject's back. Blindfolded subjects had to move their arm and finger and when the camera recorded the target object, they received a tactile stimulus on their back. The task was to estimate the distance of the target object in three sessions with 48 h between sessions. Subjects did not receive any feedback about their performance. They were divided into two groups: one should concentrate on the posture of their arms and try to triangulate the location of the object when estimating the distance (PA, for 'proximal attention'). The other group was told to concentrate on the distal object, thus relying on an "intuitive sense" of the location of the object when estimating the distance (DA, for 'distal attention').

The rationale for interpreting the results was the following. The basic idea was that one can read off from the subject's performance whether there is distal attribution (in the authors' sense of perceived objects in space) or mere triangulation (i.e. inference) going on in sensory substitution. The authors assumed that in case the performance of the subjects would not change over three sessions this would provide evidence for the cognitive character of distal attribution, since subjects received no feedback about the accuracy of their triangulations and therefore would show no changes. This is exactly the pattern of results that was observed for the PA group: they neither performed more accurately nor did the variability in their judgments reduce over time. Subjects of the DA group, in contrast, showed an increase in accuracy and a decrease in individual variability. The authors interpret these findings as evidence for the view that members of the DA group developed an increasingly concrete perception of the target object and its location, which allowed them to better estimate the distance of the object.

It seems clear that both groups did not rely on the same means to estimate the distance of the target, since their patterns of results differ significantly. Furthermore, it is clear from the results that the estimates of the DA group became more definite, which is shown by the decrease of individual variability. Thus, it seems that the DA group did not use a cognitive strategy to estimate the distance of the object as the PA group apparently did. However, the question remains whether this shows that subjects of the DA group literally *perceived* the object.

Everything hinges on the meaning of 'perception' that is in play here. If one thinks, as Siegle and Warren seem to do, that there are only two reasonable options, namely either an explicit triangulation or an intuitive sense of the location of the object, and the latter means that subjects perceive the object, one might conclude from the results that members of the DA group had (or developed) a perception of the object. However, one should not forget that the participants knew in advance that there was an object present that caused their tactile sensations. Hence, it might be the case that they merely *imagined* an object at some plausible distance and that these imaginings became more definite in the course of the experiment, something that would have been easily possible given their background knowledge about the situation. In effect, I question the simple identification of 'having an intuitive sense' with 'perceiving' that the authors employ.

I cannot do more here than suggest some argumentative strands one could take to resolve the issue. To get clear about the exact border between perception and imagination would itself require a book. Therefore, I only propose some steps toward the one or the other direction.

Maybe the prospects of arguing for the perceptual character of the resulting experience are not too bad. At the most coarse-grained level, mental imagery is taken as a perception-like experience "triggered by something other than the appropriate external stimulus" (Gendler 2013). Taken seriously, one might hold that what differs between perception and imagination is just the existence of a causal chain between the perceived object and the perception of the object, something that is lacking in the case of imagination but in place in the case of perception. On this basis, one might argue that Siegle and Warren indeed manage to show that there is perceptual experience in usage of SSD, since there is a causal chain from the perceived object to the perception of the object in their experiment.

However, the question now is: Is the pattern of tactile sensation on the subject's back an "*appropriate* stimulus" for the resulting experience, as the above given characterisation demands? Defenders of the perceptual view might say 'Yes', since it is caused by the experienced object and forms the basis of the experience, thus pointing toward the perceptual character of the resulting experience. However, one might also answer 'No' to the question, since the stimulus only gives rise to perceptions of pressure on the subject's back. The putative experience of an object somewhere in front, however, might be based on these perceptions and the background knowledge about the whole situation including the functioning of the SSD, resulting in an imagination of the object somewhere in front. On this latter proposal, the stimulus would not be appropriate; it would at most be a trigger of the imagination.

One might also argue that both answers are right, with regard to different points in time of the experiment. Maybe at the beginning there was just an 'imagined hypothesis' of where the object would be that was based on the tactile stimulations and the proprioceptive perception of the moving arm. However, with practice, the situation changed and the experience eventually became more of a perception of the object. Possibly this would have been no vision-like perception but more like a pure spatial perception of a quite concrete object in front of the subject (cf. Martin and Le Corre 2015).

One could also argue in a substantially different way based on the conviction that perception and imagination differ in something other, or in more than the presence of the appropriate stimulus, e.g. in the phenomenological character, the notorious vividness that imaginations are said to lack compared to perceptions. It would require more empirical work to assess the extent of the vividness, but primarily it would afford a measure of vividness and some means to decide the borderline between imagination and perception on that measure, something that seems hard to develop.

As I said before, I cannot do more than highlight these issues and propose some possible options here. It should be clear that a sceptic about the perceptual nature of the experience resulting from using SSDs need not be fully convinced by the cited results.

In sum, what is the moral of the discussion of these experimental results, what do they show? First, both studies found some form of distal attribution, in the sense that subjects grasp their sensations as caused by something external, given the right circumstances, even when they are ignorant about the functioning of the device they are using. This positive result nicely adds up with the anecdotal evidence in the literature. Second, the question about how exactly to characterize the distal attribution that is present in these cases is not entirely settled. There is reason to doubt that it is only a matter of consciously drawing inferences from tactile/auditory sensations to their putative sources. Although this might be what is happening in initial stages of these experiments, the differences in the result patterns of the second study would not be intelligible if this was all to be said here: if both groups of participants would rely merely on inferences they should show the same result patterns. However, even if it is not just inferences, it still has to be shown that it is perception, since imagination is another plausible candidate that is not taken into account by the authors.

Finally, taking together all the evidence I have presented, is there actually spatial perception of a distant object proper in sensory substitution? I think there are good reasons to think there is, although there is no complete confirmation for that yet. Practically all reports of subjects using SSDs state that the sensations of the substituting modality become transparent after the subject is trained so that these people do not describe their experience as being a perception in the substituting modality. Wouldn't the reports sound different in case there was only auditory/tactile perception plus inferences on the side of the subjects? In addition, the second study on distal attribution I presented indicates that there is more to the story than a cognitive process. Cognitive inferences may play an important role in learning to use these devices and in using them, i.e. in learning to know where what is, based on the deliverances of the device. However, this alone does not mean that there is no perception of objects in space going on in these cases. Our normal everyday perceptual experience is also loaded with background beliefs, expectations etc., which is seen in such ubiquitous phenomena as amodal perception, i.e. the perception of parts or features of things that have no current causal impact on sensory systems (Briscoe 2011). Perception is not constituted alone by what hits the surface of the body somewhere. Of course, it is likely not the case that the perception acquired with SSDs is as rich as normal visual perception. I already said that the affective dimension is lacking, colours are not perceived as well, and the resulting perception of objects and locations probably is not as precise as in normal perception. However, all this does not rule out that what is going on in these cases is not of perceptual character. Moreover, if we take for granted the reports by PF in Ward and Meijer (2010), practically the only reported case of a person that has used an SSD over a long period, we have good reason to regard the resulting experience as perception. The same holds for the report in Heil (1987) about the infants that seem to use the device quite naturally. At least it is enough to shift the burden of proof to those who think that there is merely perception in the substituting modality plus inferences that enables the subjects to know about their environment.

3.1.4.2 Which account best accommodates the results?

Let me turn to the second question of how to interpret the observations and results concerning sensory substitution. What constitutive relationship may be shown by cases of sensory substitution? I want to consider three; the sensorimotor account, an Evansian, and a Schelerian version of the dependency thesis.

O'Regan and Noë (2001) argue that sensory substitution cases can show that what constitutes perceptual space is knowledge of sensorimotor contingencies. They rely heavily on one particular observation that is often made in the literature about sensory substitution. The observation in question is that subjects are only able to learn to perceive with these devices when they can actively manipulate the camera (Bach-y-Rita 1984; Bach-y-Rita et al. 2003; Noë 2004, pp. 26–7, pp. 110–6; Hurley and Noë 2003, p. 142; Hurley 1998, pp. 344–6).[131]

> This important point constitutes an empirical verification of the mainstay of the present theory of visual experience, namely, that seeing constitutes the ability to actively modify sensory impressions in certain law obeying ways. (O'Regan and Noë 2001, p. 958; cf. Hurley and Noë 2003, p. 145)

Now, is that right? I do not think so. For one thing, one has to distinguish between developmental and constitutive questions. It might be right that subjects are only capable of perceiving with SSDs in case they are able actively to determine the movements of the camera. However, this may be just a necessary condition on a successful development. This does not show that the resulting perception is constituted by the thereby acquired knowledge of how sensations change with movements.

Second, there is empirical evidence that subjects acquire sensorimotor knowledge in the sense that the defenders of the approach have in mind without thereby acquiring perception of a distant object. One of the main results of the study by Auvray et al. (2005) that I discussed above is that many subjects noticed that the sounds they heard were linked to their movements in lawful ways (p. 518). However, this was the case without these subjects perceiving or even judging that their sensations originate in a distant source. They just observed that their movements determine the pattern of sounds they heard but no spatial impression was immediately linked to that (cf. Grush 2007, p. 413). This seems like an empirical refutation of the thesis that sensorimotor knowledge is what constitutes perceptual space, a thesis defended most notably by O'Regan and Noë. Accordingly, the authors in the discussed study (re-)formulate their central statement as a weaker, developmental thesis, stating that the realization of a sensorimotor coupling is a necessary, although

131 Macpherson (2013) questions the claim that a necessity of active movement has been proven yet.

not sufficient condition for the development of perceptual space (Auvray et al. 2005, pp. 519–20). Note that one defender of the sensorimotor view, Kevin O'Regan, is a co-author of this study.

A third and final observation that questions the sensorimotor account comes from the second part of the study by Siegle and Warren (2010), already discussed above. Remember that they used a small camera that was mounted on the index finger of the subjects. In the second part of their experiment, they tested the performance of subjects in estimating the distance of the target object after either they were turned around or the camera was mounted onto the index finger of the other hand. The results clearly indicate that there was no difference in performance between the last trials of the first part and the first trials of the second part in neither of the different conditions (rotation vs. arm transfer). Subjects did not have to 'adapt' to the new situation although the pattern of sensorimotor contingencies had of course been completely altered.[132] The authors rightly highlight (p. 221) that these results are not consistent with the sensorimotor view as suggested in O'Regan and Noë (2001). There, it is proposed there that the relevant sensorimotor laws that are taken to be constitutive of perceptual space are explicitly described at a joint-specific level (p. 956). However, knowledge about such laws for wearing the camera on one finger, or in one orientation, is of no use when wearing it on the other finger, or in another orientation, one might argue. How then is it possible that subjects immediately after the transfer showed the same pattern of performance than before? It seems that sensorimotor laws at this level of specificity cannot explain this observation. In other words, subjects perceived the location of the target object with the same level of accuracy although they were ignorant of the laws of the couplings between their limb movements and sensations, an observation that is in conflict with the central tenet of the sensorimotor view. To accommodate these findings, defenders of the sensorimotor view would have to alter their conception of the relevant sensorimotor laws, e.g. to formulate them on a higher level of abstraction.

132 This was also already observed e.g. in Bach-y-Rita (2004).

I think we ought to conclude from these three points that the sensorimotor view as O'Regan and Noë formulate it cannot explain the experimental findings and therefore it appears as an inappropriate explanation of perceptual space.

What about the remaining possibilities? Is it plausible that these cases can be used as evidence for a view that spatial perception constitutively depends on action, in either an Evansian or Schelerian fashion?

Rick Grush (2000; 2007) has argued that sensory substitution cases speak in favour of an Evansian approach. He asks what difference holds between someone who can use an SSD to perceive the environment and someone who has just put on the device. On a certain level of description, there is no difference at all between the two cases: given that both are in the same environment, situated and moving the same way, the sensations they have are identical. For both subjects would receive the same sounds or pressures as they walk around. However, their perceptions would differ dramatically. Whereas the trained subject would perceive objects at certain locations and the shape of the environment, the untrained subject would merely perceive continuously changing patterns of sounds.[133]

An answer to the question about the constitutive conditions on spatial perception can be reached in comparing both cases. The most obvious difference between the two cases is that, in contrast to the untrained subject, the trained subject is able to use the deliverances of the SSD to guide her bodily actions with respect to distant objects (Auvray et al. 2007; Segond et al. 2005; Levy-Tzedek et al. 2012; Proulx et al. 2008; Renier and Volder 2010; cf. Bach-y-Rita 2004). As we saw, this is not a matter of reasoning based on sensory patterns and background knowledge; rather, their perceptions can serve certain functional roles that the perceptions of untrained subjects cannot.

133 Maybe one should rather say that the trained subject enjoys a perceptual dimension that the untrained lacks. The trained subject arguably perceives sounds, too. In addition, she perceives objects etc. whereas the untrained subjects lacks this dimension. For the trained one the experience might be somewhat similar to touch where one can perceive the pressure on the skin and the spatial features of the object one touches depending on what one focuses one's attention on.

Whereas the trained subject can e.g. grasp an object based on what she perceives, the untrained subject cannot do so. Therefore, the practical dispositions that are connected with the sensations of the trained subject account for her spatial perception.

Against this proposal, one might formulate the following objection. One might argue that it is no surprise that trained subjects can act based on what they perceive with the device, since they perceive objects, the environment and their spatial features. That they can act in this way is rather like a sign that they have spatial perception than evidence that action is a constitutive condition on spatial perception. The objection thus states that the possibility of acting with respect to a perceived object is not more than an evidential condition on having spatial perception, meaning that bodily goal-directed action implies spatial perception. If this were right, the direction of explanation as it is proposed by accounts that take spatial perception as constitutively dependent on action would be wrong.[134] The rationale behind this objection is the following. What is obvious from cases like sensory substitution is that there is a co-occurrence of spatial perception and bodily goal-directed actions to be observed. Now, what one certainly can say is that goal-directed actions, i.e. voluntary bodily movements toward some external object require the subject to somehow know or be aware of the location of the object relative to itself, something that is accomplished by perception.[135] Hence, the fact that subjects using SSDs can perform such actions and that this is not based on mere reasoning is indicative of the fact that they enjoy spatial perception. However, this does not show that action is a constitutive condition on spatial perception. How could action be constitutive when it itself depends on the presence of spatial perception? The presence of spatial perception thus needs a different explanation.

134 Note that this objection is applicable to the whole kind of approach that takes action as a constitutive condition on spatial perception. I already noted this issue above in relation to the short presentation of Peacocke's (1983) position and how it is different from Evans's.

135 Of course, a given location relative to oneself might be known in a different way. However, since I deal with spatial perception here, I will confine myself to the case where an object's location is known by perceiving the object.

To this line of thought, one might respond in the following way. An exercise of some goal-directed bodily action might require a perception of the object that specifies its location in relation to the subject. This, however, does not show that the constitutive claim that is defended here is false. First, two things might be mutually constitutively dependent on each other. I can see no reason in the concept of constitutive dependence that precludes this possibility.[136] Second, even if this were not the case, the critique would not be successful. The fact that goal-directed bodily action depends on spatial perception does not preclude that spatial perception is dependent on the *capability* to act. The objection is phrased in terms of token actions whereas the hypothesis in question is framed in terms of abilities. Thus, it is argued here that the ability to act based on the deliverances of the device is what constitutes the spatial contents of the perceptions.

Finally, I want to ask whether sensory substitution cases can serve as a means to decide between an Evansian and a Schelerian account. Remember that while for both accounts dispositions to act are crucial, they are so in different ways. Whereas on Evans's account the factual presence of dispositions that are directly linked to perceptual information is crucial, on Scheler's account it is some grasp of these dispositions to move that is constitutive.[137]

What would have to be observed in sensory substitution cases for these cases to be a means to decide between the two varieties? To show that sensory substitution cases support the Schelerian rather than the Evansian view, one would have to make out whether the difference between the trained and the untrained subject lies in the former's ability to orient toward an object based on his perceptions alone, or whether it lies in his grasp of this ability. In other words: Is spatial perception conditional on the bare ability to interact with distal sources of one's perception, or is it conditional on a sense of one's abilities to interact with the distal sources of one's perception? To decide experimentally between the two alternatives one would accordingly have to be able

136 Cf. the discussion of the notion of the constitutive relation in section 2.1.2 above.
137 Cf. the interpretation of inverted vision cases in section 3.1.3 above.

to differentiate between the case where a subject has spatial perception because he is able to orient toward a distant object and the case where a subject enjoys spatial perception due to his grasp that he is able to turn towards a distant object.

One might be tempted to use one observation that is quite well established in the literature to argue in favour of a Schelerian instead of an Evansian view. The relevant observation is that successful behaviour based on SSDs is possible *without* the presence of spatial perception (Auvray et al. 2007; Segond et al. 2005). From this, one might try to build an argument to the effect that spatial perception constitutively depends on some grasp of action possibilities rather than the mere factual presence of these possibilities. The rationale of this line of thought would be to show that while all conditions for an Evansian approach are satisfied, there is no spatial perception. In the next step one would then argue for the fact that subjects acquire some grasp of their capabilities (which seems probable due to training) and that this grasp is what grounds spatial perception.

However, this line of thought would be non-persuasive since it ignores another fact that I discussed at large before: namely, that in the initial stages of the experiment, subjects arguably are able to use cognitive strategies to infer the location and shape of things based on their auditory or tactual perceptions. Assuming this is the case, the conditions for Evans's account are not met since a crucial ingredient of Evans's approach is the absence of any thinking or calculating of how one needs to move e.g. to turn towards the object (Evans 1982, p. 155). In effect, in these initial stages of using SSDs there are no behavioural dispositions linked to the perceptual information in the direct way that Evans proposes. Thus, the argument would fail.

I am sceptical whether present empirical evidence from work on SSDs can be used to decide between the alternatives. The experimental paradigms that are used hitherto are not fine-grained enough to provide compelling reasons for one or the other approach. In Ch. 4.1, I will present an account that is inspired by Scheler's basic idea an argue that it is more appropriate than Evans's.

3.1.4.3 Summary

In sum, sensory substitution cases provide good evidence for the general claim that dispositions to act are constitutive of spatial perception. They are especially valuable since in these cases one can directly compare subjects that enjoy perceptual space and subjects that do not, although they receive the same kind of input, i.e. the same kind of proximal stimulation. It has been shown that the key difference arguably is the presence of action dispositions on the side of subjects who are able to perceive their vicinity with the device. What is also obvious from experiments using SSDs is that knowledge about sensorimotor contingencies, i.e. knowledge about how sensations vary with movements systematically is not enough to ground perceptual space. Instead, these cases strongly speak in favour of theories that take dispositions to act as the essential explanatory element, even if they cannot be used to adjudicate decisively between the specific versions of such a theory.

3.2 Empirical Challenge: The dual visual systems hypothesis

In the second chapter of this part, I want to consider a possible empirical threat to the idea that action is part of the constitutive conditions on spatial perception. There is a wide-ranging body of evidence showing that two different neural systems process visual information in the human brain according to different functions: the ventral stream subserves perception and the dorsal stream subserves visually-guided action (Milner and Goodale 1995; 2006; Jeannerod 1997). The approach is based on neuropsychological studies as well as experimental dissociation, including visual illusions. An important implication of the view is that what you consciously perceive is not what guides your actions with respect to what you see. The information that is conveyed by perceptual experience is not suitable for the control of actions. Perceptual experience only serves such tasks as object recognition, or memory. When it comes to visually-guided actions, perceptual experience is useless. These actions rely on the processing of the other, dorsal system.

Given this picture of vision, it appears that any account that proposes a constitutive dependence of perception on action is called into question. If it were true that visual experience is not linked to visual-based actions, how could it constitutively depend on the latter? I will investigate if and how the dual visual stream approach (for short: 2V) challenges the constitutive accounts discussed in Ch. 2.2 and 2.3 above. I will argue that 2V can be reconciled with all of the constitutive approaches.

3.2.1 The challenge that 2V poses

The dual visual systems hypothesis (2V) has been put forward by the neuroscientists Melvin Goodale and David Milner.[138] The core idea is that visual information is processed in different areas of the brain relative to the function it serves. It is claimed that there are two anatomically and functionally separated systems for dealing with visual information: the vision-for-action system and the vision-for-perception system. Based on the location of the brain regions involved, the two systems are also identified as the dorsal as opposed to the ventral processing stream, respectively.

A key claim of 2V is that the information that is present in conscious visual experience of the environment is *not* the same as the information that is the basis of visually-based interaction with the world. This means that the visuomotor control of bodily actions is not due to one's conscious perception of

138 See Goodale and Milner (1992, 1995, 2008), Goodale et al. (1991), and Goodale (2011); see also Jeannerod (1997). 2V is a development of Ungerleider and Mishkin's (1982) earlier distinction between "what and where" pathways based on primate work. My discussion focuses on visual space. There is evidence that the general framework of the approach can be applied to other sense modalities: Rauschecker (1998); Belin and Zatorre (2000); Dijkerman and de Haan (2007).

the world but due to a phylogenetically more ancient subsystem of the brain whose processing of visual information does not contribute to visual experience.[139] In the ventral stream, which is held to subserve conscious perception, the information is processed so as to make possible identification, re-identification, and classification of objects. Here is where constant object properties such as size, shape, colour, and location are processed, relative to the surrounding environment, in a so-called 'scene-based frame of reference'. In contrast, the dorsal stream which subserves visually guided action codes information that is used for the control of bodily actions in an (effector-specific) egocentric frame of reference. No identification or classification of the object is accomplished in this system (you do not necessarily know what kind of object is before you to grasp it) and the information is not stored in memory (you need not remember in what way you'd have to move to grasp some object you saw minutes before).

Based on these central claims of 2V it is possible to build an argument against constitutive views. The argument would go like this:

The first step is the observation that 2V questions an intuitively plausible picture about the connection between perception and action. Here is a quote from Brian O'Shaughnessy that nicely illustrates the intuitive picture: When one wants to touch something (in this case, the target is a cross on a sheet of paper) "[...] one keeps *looking* as one guides the finger, and does so right up until the moment in which the finger contacts the cross, and the reason surely is that sight is continually informing one as to where in one's visual field one is to move one's visible physical finger" (O'Shaughnessy 1992, p. 233). O'Shaughnessy is committed to the idea that visually-based actions are guided by perceptual experience.[140] Andy Clark has called this the assumption of Experience-Based Control (EBC):

---

139 However, see Matthen (2005), pp. 304–6 who argues that the dorsal stream contributes to a "feeling of presence" of seen objects.

140 One can reason about the modal status of this claim: is conscious experience only typically involved in these acts or is it necessary for those acts to be possible? The latter thesis is considered (and rejected) by John Campbell in Campbell (2002), pp. 48–57; cf. Campbell (2003).

> Conscious visual experience presents the world to the subject in a richly textured way, a way that presents fine detail [...] and that is, in virtue of this richness, especially apt for, and typically utilized in, the control and guidance of fine-tuned, real-world activity. (Clark 2001, p. 496)

Now, in light of 2V this picture cannot be maintained. We saw that on 2V it is not conscious visual experience that guides our bodily actions toward objects but some non-conscious subsystem that links visual input to motor output without consciousness. EBC thus has to be rejected.

The second step of the argument is to claim that the constitutive view essentially involves something like EBC. If this was established it would be clear that the constitutive dependence was impossible to hold. Clark (2001), citing Evans, Peacocke, Cussins, and Grush, claims that there is such an implication between theories that posit a constitutive link between perception and action and EBC:

> "[EBC] is equally implicated both in a certain intuitive picture of the functional role of conscious visual experience (roughly, that we use conscious seeing to guide fine motor activity) and in any philosophical account that seeks to fix the contents of perceptual experience (whether conceptualized or not) by invoking direct links with action." (p. 499)

Thus, it seems that the argument against the constitutive connection between action and perception is complete: 2V shows that EBC is wrong; EBC is an essential part of theories that posit a constitutive connection between action and perception; therefore, the view that perception constitutively depends on action cannot be sustained (at least not in the way that Evans and others have proposed).

This is the challenge that 2V poses to the constitutive view.[141] In the following section, I will discuss the evidence put forward in favour of 2V in some detail, so that we can judge whether the constitutive views can respond to this challenge.

141 Briscoe (2009) has an interesting critical discussion of Clark's reasoning. I do not think that this bears on my argument here. In discuss his account in the Appendix to this chapter.

3.2.2 Considering the evidence for 2V

Two main sources of evidence have been used to support 2V. On the one hand, there is the observation of a double dissociation between two pathologies, optic ataxia and visual form agnosia.[142] The former is a disorder that is caused by damage to parts of the dorsal stream and results in patient's inability to interact with visually perceived objects while being able to describe the scene verbally (Milner and Goodale 2008). The latter is manifested in an inability to recognize or describe seen objects while being capable of correctly interacting with them. These deficits are due to lesions to parts of the ventral stream. As is obvious, the two pathologies are marked by reverse patterns of deficits, thus allowing one to conclude that both tasks, conscious visual perception and visually guided action, are subserved by dissociable brain modules.

The second main source of evidence for the theory comes from experimental dissociation studies. These studies show that certain illusions are only present in visual perception while visually-based actions remain unaffected from the illusions (Goodale and Haffenden 1998; Goodale and Humphrey 1998; Aglioti et al. 1995; Gentilucci et al. 1996). The most famous of these illusions is the Ebbinghaus Illusion. The illusion consists in seeing two discs of the same size as being of different sizes due to the difference in size of surround discs. The disc that is surrounded by larger discs is seen smaller than the one that is surrounded by smaller discs. However, this size illusion is barely reflected in the size of grip aperture when one reaches for the discs. That is, visually-guided grasping is apparently not susceptible to the illusion. This in turn implies that visually-guided movements are not based on conscious visual experience.

Let me turn to the evidence in favour of 2V that seems most challenging for the constitutive view. The first one is optic ataxia, caused by lesions to the dorsal stream. The second one consists in the dissociation studies that show that illusions do not have the same effect on action guidance as they have on

142 For details on the idea of double dissociation, its history and possible problems cf. Davies (2010). Davies argues that double dissociations are part of arguments to the best, not to the only possible explanation.

perception. I will present both bodies of evidence in turn and consider whether they present a threat to the constitutive views.

3.2.2.1 Optic Ataxia

Patients suffering from this disease are impaired in their visually-based actions. Although they can identify objects and overall accurately report what they see, i.e. visual perception is intact, these patients cannot appropriately interact with objects they perceive. This obviously seems to be a threat for the view that what determines the contents of visual perception are links to action: if action is not possible on the basis of perception, the constitutive claim is arguably false.

However, one can say several things in response to that challenge. First, these patients are not fully incapable of acting on what they see: it is not the case that they can perfectly describe the scene around them but when asked to grasp an object have no clue how to move. What is predominantly disrupted in these cases is the capability of the subjects for fast, fluent, fine-grained control of their action based on vision (Pisella et al. 2000; Briscoe 2009). Thus, the deficits are more related to time and accuracy but not to the general capability to act on what one sees. This is also confirmed by studies imposing a delay between perceiving the target objects and action onset. The somewhat surprising finding is that these patients are better at interacting with remembered objects than with currently seen ones (Milner et al. 2001). Since, according to 2V, visual memory is fed by the ventral stream (the vision for perception system) whereas the representations of the dorsal stream are only short-lasting, i.e. they have vanished at the time of action onset in these cases, these delay studies show that, despite the dissociations, there must be some connection between the perceptual system and action.

Second, even if it were the case that optic ataxia would present a more substantive dissociation between perception and action than it actually does, the constitutive view does not necessarily fail. Regarding the views that posit a constitutive dependence on dispositions to act, it is important to recognize that to have dispositions to act does not mean that one is able to actualize the movements one is disposed to make in every case. That is, there might be other conditions on the successful performance of the movements one is disposed to make that are not met. Thus, one might explain the case of optic ataxia by arguing that those affected are indeed disposed to act on what they see in various ways but that the disease undercuts the conditions that have to be satisfied for the subjects to be able to actually perform their actions.

A similar argument is possible for proponents of the sensorimotor knowledge approach. The fact that patients suffering from optic ataxia cannot accurately guide their grasping actions at what they perceive is not in conflict with the claim that the knowledge about how visual patterns varies with movement constitutes spatial perception.[143] The know-how is a learned association between movements of the subject and the resulting patterns of changes in sensory patterns. Why should it be processed in a system that grounds online control of visual-based actions?

I think these three points show that the case of optic ataxia presents no empirical refutation of constitutive views. It would present only a threat to theories that take the actual performance of those kinds of movements that are impossible for patients suffering from optic ataxia as constitutive for spatial perception. Hence, the only thing that is shown by optic ataxia is that spatial perception does not constitutively depend on the actual exercise of those highly automated, non-consciously controlled bodily acts. However, none of the constitutive theories outlined above hinges on such a claim. Dispositions to act or a grasp of own practical possibilities might as well be present as knowledge about how sensational patterns alter with movement even when the actual execution of these movements is impossible. Therefore, optic ataxia does not conflict with the constitutive views that are considered here.

Let me turn to the second source of evidence that may seem problematic for constitutive views.

143 Brogaard (2011) has argued that 2V challenges enactive accounts such as Noë's since it follows from the view that dorsal processing never contributes to conscious perception. However, I cannot see why the relevant know-how about sensorimotor patterns has to be part of the dorsal system. I will shortly discuss her argument in the Appendix, too.

3.2.2.2 Experimental dissociation studies

The key finding here is that visually-guided action is not as susceptible to visual illusions as visual perception is. This seems to speak against the idea that the spatial contents of visual perception are determined by the actions of the subject in any way. At first glance, one would expect that if the constitutive view was right, the properties of the actions that I am disposed to determine the spatial properties of perceptual space for every token perception in the same way: seeing something to the right means (in part) being disposed to turn a certain way (e.g. if one wanted to look at the object) for every instance of seeing something at that particular (egocentric) location. Analogously, one would expect that a given perceptual size of something seen is linked to a distinctive way of acting, e.g. of opening the thumb and index finger.

The experimental dissociation studies show that this is not necessarily the case. Experiments with graspable versions of visual illusions such as the Ebbinghaus Illusion show that one does see things as of different sizes although one reacts in the same way to them in terms of grip scaling. Thus perceptual size and 'motoric size' do not always vary in the same way, something that undermines the idea that the spatial content of perception is determined by the features of the actions one is disposed to.

How could proponents of the constitutive view respond to that challenge? The first option certainly is to question the results of the studies by highlighting that there is some evidence that these illusions have a similar effect on action as they have on perception (Franz and Gegenfurtner 2008). However, this has only been shown for the Ebbinghaus Illusion but not for other illusions that show the same dissociation (Nanay 2015, note 2). So it seems that we have to accept that there are possible situations where perceived size and 'motoric' size come apart.

Another possible reply could be to insist that the *explanandum* of constitutive views is egocentric spatial perception, i.e. the perception of objects and features being located at various distances and directions with reference to the perceiving subject. Thus, it is about perceiving locations, not shapes or sizes, on which the experiments focus. One might then propose that constitutive views need not include claims about a constitutive relation between internal spatial properties of perceived objects (size, shape) and action and that therefore the evidence from illusion studies is not relevant.

However, there are two problems with this kind of reply: First, this response would limit the scope of the constitutive views substantially. Second, the dissociation is not limited to cases of shape/size perception as is shown by Króliczak et al. (2006) who use the hollow-face illusion to show that perceived location and the location at which an act is targeted can come apart, too.

What, then, can proponents of either of the kinds of constitutive views say in response to this challenge?

Proponents of the capability approaches can argue based on a refutation of a condition on which the challenge hinges: namely that the features of disposed actions determine the spatial contents of perception entirely alone. This is a claim to which proponents of the constitutive view are not committed. The accounts we looked at claimed that the links to dispositions to act are *partially* constitutive. On views of the first kind, the dispositions to act are taken to be linked to incoming sensory information. It is not claimed that dispositions fully determine spatial content alone. Thus, it is possible that the illusions are due to the processing of input signals, independent of their links to action.

Besides that, it is also not the case the we cannot act at all on the basis of these perceptual illusions. This is most clearly shown by the ability to goal-directed actions of patients suffering from optic ataxia, i.e. patients whose dorsal processing is distorted and who have to rely on the processing of the ventral stream, and therefore their conscious perception (see above). The actions that are based on perceptual experience alone are not as fast and flexible as those controlled by the dorsal stream – but this does not show that perceptual experience is not connected to the capability to act. To the contrary – given the case of optic ataxia patients there has to be such a connection.

The situation for sensorimotor approaches is more difficult. Remember that the core claim there is that practical knowledge about sensorimotor contingencies constitutes perception. It follows from this thesis that seeing the central discs of the Ebbinghaus Illusion as of a different size means that the contents of the sensorimotor knowledge about the central discs differs. But then it may be difficult to explain why the actual motor performance is the same for movements towards both discs.

In response to this point, one might argue that grasping movements are not crucial here since they do not have an immediate effect on the sensory patterns. What is relevant are e.g. eye-movements (Noë 2004, p. 90). However, this might be problematic too: Suppose you only see the lines of a Müller-Lyer Illusion first, without the arrows on their ends. You just see two lines of equal length. Given the sensorimotor approach, this is due to the same sensorimotor knowledge about how the sensory patterns caused by the two lines would change with movement. Now, suppose the arrows at the ends of each line are added, resulting in the impression that the lines are not of the same length anymore. Is it plausible to assume that the sensorimotor profile of the arrows is different from the one of the lines before? Both are in fact of the same length. And why couldn't one at least learn the relevant knowledge by repeated encounters which would let the illusion finally disappear? These considerations cast some doubt on the sensorimotor approach and its ability to accommodate the differences in perceiving and acting on perceptual illusions. Surely, they are far from undermining the sensorimotor account, but at least they put some pressure on the side of proponents to explain the differences. The easiest way of doing this certainly would be to highlight the different context that is created by turning the lines into arrows.

I have not yet said anything about Hurley's approach yet. I think that for this case one has to consider the issue more generally. Hurley's account is pitched at the level of particular cases, not on the general level of the relation between the capabilities for action and perception. For 2V to be a threat to Hurley's picture, it would have to rule out the possibility that ventral processing is in part determined by motor signals, since this would be exactly a case of non-instrumental dependence in Hurley's sense. However, this possibility seems not to be ruled out by 2V; in fact, 2V does not seem committed either way on this front. The explanations given by 2V are unidirectional: from retinal input to perception or perceptually-guided action. There is no claim made about possible links in the other direction: from output signals to input signals. It is just claimed that visual input is separated according to the end it serves, *viz.* perception or action. It is not claimed that the processing of input signals is not affected by output signals in any way. Hence, for this general reason, 2V does not present a threat to the third kind of views at all since proponents of 2V are not committed to denying the possibility of a non-instrumental dependence between input and motor signals.

Furthermore, there might be some way of reconciling both approaches based on evidence that supports the claim that there are common representations of spatial features that are the basis for the processing of both the ventral and the dorsal stream (Christiansen et al. 2014). Thus, in terms of processing stages, before the streams actually separate there is a stage including a common representation. This could in principle be some point where motor signals fed back have an effect on the processing of visual input signals.

In sum, none of the three kinds of approaches is severely challenged by the results of these studies. The capability and sensorimotor approaches (the latter with more effort, though) can explain the results within their framework whereas Hurley's approach seems not challenged by these results at all. It follows that neither source of evidence for 2V challenges any of the constitutive views. This suggests that there is a way of reconciling both approaches. In the final section, I want to propose a way of doing so.

3.2.3 United, not divided

I have argued that it cannot be part of 2V that perceptual experience cannot guide bodily actions at all. Recently, Milner and Goodale (2008) themselves have admitted that ventral processing has a part to play in visually-based action. They emphasize the distinction between planning on the one hand and programming of visually based actions. The latter process that is subserved by dorsal processing means the "pre-specification of movement parameters". The former process refers to the selection of an appropriate action. An example for this selection given by the authors is the selection of where to grasp an object. The selecting process is said to be subserved by the ventral system, i.e. perception. It takes into account the kind of the target object, but for the planning to be possible, it must also incorporate the spatial properties of the seen object, both shape and size as well as location relative to the subject. Thus, a direct connection between the spatial contents of perceptual space and *planned*, *intentional* bodily actions seems compatible with 2V. Milner and Goodale's reformulation is similar to Clark's (2001) hypothesis of "Experience-Based Selection" (EBS), according to which experience is linked to "the reason-and-memory-based selection of action".

I want to make three comments about this proposal. For one thing, far from undermining the constitutive connection between action and spatial perception, this proposal is reminiscent of a Schelerian approach to the constitutive view in that perceptual space is linked to some grasp of one's practical possibilities. One way of cashing out this idea is by alluding to planning processes (as discussed e.g. by Glover 2004). In planning some action one has to have some sense of what one can do and what one cannot do.

Second, we have to keep in mind what we have learned from the case of optic ataxia. Despite lesions in the dorsal stream, these patients are actually able to act based on what they see, they are just slower. Hence, it seems that, at least when the processing of the dorsal stream is distorted, the results of ventral processing can control bodily actions as well. This process might be slower due to action-unrelated information that is processed (Briscoe 2009). In any case, it is shown that there *is* some link from perception to action.

The third comment I want to make refers to a seemingly important property of that link. It is about directness. Clark argues that a direct link between perception and action cannot be sustained but only an indirect one between perception and reasoning about what to do. On the one hand, claims about directness should only be made based upon a clear conception of what directness means in this case. This is something that Clark does not provide. On the other hand, I think one could as well claim, based on the evidence from 2V, that there *is* a direct link, although not from perception to action in general but to *intentional* action. In other words, I suggest that it is not necessary to say that experience is merely indirectly linked to action. One can also hold that it is directly linked to *intentional* action. Were perception only related to the process of reasoning and selecting for action, as Clark proposes, those subjects that have to rely exclusively on their vision for perception system in acting would not be able to actually move purposefully based on what they see. This, however, conflicts with the description of their behaviour, at least as given in Pisella et al. (2000): they have no 'automatic pilot', but this does not prevent them from 'flying manually', as it were.[144]

I think that we can conclude that 2V is not a serious threat to the view that action is a constitutive condition on spatial perception. Contrary to first appearances, 2V does not imply a wide gap between perception and action in general. From the three kinds of approaches that I have discussed, only the second that is based on sensorimotor knowledge might be taken as having some difficulties with dealing with 2V. The other two sorts of approaches can be reconciled with the evidence.

3.2.4 Summary of Part 3 on empirical evidence

In this part, I have presented various sources of empirical evidence that speaks in favour of the constitutive view and one piece of evidence that has been taken to threaten the view. At first sight, all mentioned empirical results are highly relevant for the investigation of the connection between action and spatial perception.

The 'Proffitt studies' seem to show a surprisingly direct and instantaneous impact of action plans on perceiving the environment, suggesting that the perceived spatial properties of the vicinity, like slants or distances of objects, vary systematically with one's plans to act and some intuitive sense of what it would take to execute the plans. The easier a planned act looks to be achievable the nearer/flatter is the target object/slant perceived, for example.

144 Rossetti et al. (2010) argue that the impairments of patients with optic ataxia are confined to bodily acts with regard to objects seen in peripheral vision but not to objects seen in central vision. The authors also highlight that these patients are not impaired in their everyday life.

The studies involving inverted vision indicate that sufficient training of interacting with the environment despite perceiving it in a fundamentally distorted way eventually lead, if not to a complete re-instantiation of the original perceptual experience, at least to a substantial change in the way the environment is perceived. Arguably, the capability to act based on visual perception has an impact on the way these perceptions are spatially structured.

Next, sensory substitution devices allow contrasting the presence and absence of spatially significant perceptions despite identical external circumstances. Of two subjects that receive the same information, only one enjoys spatial experience, but the other does not. The difference can be explained by the first being able to act on what she perceives while the other is not.

Finally, an influential view about the neural processing of visual information, claiming that visually based actions are not directly based on conscious visual experience but are subserved by a different way of processing visual information, seems to imply that action and perception are more indirectly linked than it would follow from the constitutive view. The Two Visual Systems Hypothesis seems to threaten the view that spatial perception constitutively depends on the capability to act since it denies that conscious visual perception is tightly linked to visually based action.

At first glance, all four sources of empirical evidence – be it in favour or against the constitutive view – seem to make important contributions for the question whether one or another version of the constitutive view is appropriate. However, I argued that when it comes to the details, the first impression of clarity vanishes relatively quickly. The 'Proffitt studies' face substantial issues – empirical, methodological and philosophical. In assessing the significance of the findings for the constitutive view, I argued that these cases are at most indicative of the constitutive dependence of perception on action in the sense of Hurley's non-instrumental dependence since they concern particular episodes of perception and action. However, it might also be possible that these cases exemplify only an internal instrumental dependence. Since this notion is only mentioned by Hurley but not well developed, it is hard to finally assess these cases. Therefore, the discussion of the Proffitt studies could only help in clarifying what empirical evidence has to look like to be of any use and it led to an important re-evaluation of the instrumental/constitutive distinction that is put forward by Hurley. Worthy as it is for these purposes, one has to admit that the 'Proffitt studies' could not really help to render the constitutive view more convincing.

As for the studies on inverted vision, I discussed several issues that need to be clarified when it comes to the details of the description and interpretation of these cases. There are disputes about how the resulting experiences are best described and about how to interpret the reports given. I argued that two observations about these cases are most important. First, that the subjects quickly learn to use highly distorted visual information for action guidance and second, that, given a sufficiently long exposure to inverted vision, perceptual space gets re-inverted. The second observation provides evidence for the fundamental conviction of approaches that take action as constitutive in that it shows that spatial perception is not determined by spatial properties of the retinal image alone. However, the first observation is problematic since it highlights a temporal gap between the ability for visually-guided actions and the change in the orientation of the visual field. I argued that this is especially a problem for sensorimotor approaches, since these imply that spatial perception would change as soon as the (new) sensorimotor knowledge is present. In contrast, capability approaches, and in particular Schelerian approaches, fare better in accommodating the results. I claimed that inverted vision studies could be used only to argue that action is a condition on the distinctive spatial organization of perceptions. They are not useful to argue that action is a constitutive condition on the ability of spatial perception. This is because they reveal how action has an effect on already established (if distorted) spatial perception.

The most important sort of empirical evidence consists in observations and results from the use of sensory substitution devices. As with the other evidence, the discussion brought up difficult issues, this time mainly in interpreting the resulting kind of experience as genuine spatial perception. I concluded that there are good reasons to regard the resulting experience as genuine spatial perception and that therefore the cases can be used as evidence for the question whether action is a constitutive condition on perception. I argued that the results of these studies speak against the sensorimotor knowledge approach, since the presence of sensorimotor knowledge does not lead to spatial perception. In contrast, the accounts that take dispositions to act are in a better position to deal with the experimental findings, even if it is not possible to argue successfully in favour of one of the remaining accounts.

Finally, the discussion of the Two Visual Systems Hypothesis revealed that it is not as threatening for the constitutive view as it initially seems to be. It was made clear that even within this framework there has to be some close relation between intentional acting and perceptual experience, even if some aspects of bodily acts like fast fine-grained online adjustments are not based on visual experience. I will further discuss the philosophical treatments of the Two Visual Systems Hypothesis in the appendix to this part.

In general, I think that the empirical evidence suggests that action should indeed be considered as a constitutive condition on spatial perception, even if the case is not as straightforward as it seems at first glance. Furthermore, it is possible to argue at least against one particular account, namely the sensorimotor knowledge account as developed by Alva Noë, Kevin O'Regan and others. What is left are the two versions of the view that the capability to act, either its factual presence or a basic grasp thereof, is to be taken as constitutive. To my mind, it is not possible to adjudicate between these two possibilities based on the present empirical evidence. In the following and final chapter, I will present some considerations that speak in favour of the Schelerian view. I will finally use this view as a premise for arguing in favour of the self-representational character of egocentric representation.

Appendix to Part 3: Philosophical implications of the dual visual systems approach to vision

Introduction

According to the neuroscientific dual visual systems hypothesis (2V), visual information is processed in two separate systems relative to its function. The vision-for-perception system underlies conscious visual experience, whilst the vision-for-action system underlies (unconscious) control of visually-guided actions. On the face of it, the approach challenges two widely held connections between conscious perception and action: It casts doubt on the intuitive understanding that we guide our bodily actions by conscious experience – and it questions the view that action is a constitutive condition on conscious experience. In this essay, I will criticise philosophical arguments that focus on 2V and centre on either side of the interconnection between action and perception. Some philosophers have attacked 2V, aiming to secure the action-guiding role of conscious experience. Others have relied on it in arguing against views that take action as a constitutive condition on perceptual experience.

The first section will focus on three arguments centring on the claim that perception constitutively depends on action. In particular, Briscoe (2009) argues against Clark (2001), that 2V does not challenge all constitutive views; Matthen (2005) argues that a certain feature of visual perception, namely the experience of objects as present, relies on the close connection between vision and behaviour as present in the processing of the vision-for-action system; Brogaard (2011) uses 2V to argue against one particular constitutive view, namely the sensorimotor approach to visual perception (Noë 2004). I will argue that the arguments presented by Briscoe, Matthen, and Brogaard are mistaken.

In the second section, I will criticise two arguments that aim to defend the view that we usually guide our actions by conscious experience. Wallhagen (2007) tries to defend the claim by relying on the distinction between conceptual and non-conceptual contents of conscious perceptual states. Mole's (2010; 2013) argument, in contrast, relies on the idea of an embodied demonstrative.

In sum, then, I want to show that these arguments, most of which challenge core claims made within 2V, are mistaken. Thus, this paper is a defence of 2V against philosophical attacks. However, I want to emphasise that my critique only applies to these particular ways of arguing – I do not think that 2V could not or should not be criticised from a philosophical point of view.

A brief introduction to 2V

2V has been put forward by the neuroscientists Goodale and Milner (Milner and Goodale 1992; 1995; 2008; Goodale et al. 1991; Goodale and Milner 1992; Goodale 2011; see also Jeannerod 1997). The core idea is that visual information is processed in different areas of the brain relative to the function it serves. Accordingly, there are two anatomically and functionally separated systems for dealing with visual information: the (unconscious) vision-for-action system, subserving visually-guided behaviour and the (conscious) vision-for-perception system, subserving visual perception, identification of objects and memory. A central claim of 2V is that the information that is present in conscious visual experience of the environment is *not* the same as the information that is the basis of visually-based interaction with the world. In the ventral stream, which subserves conscious perception, information is processed so as to make possible identification, re-identification, and classification of objects. Here is where constant object properties such as size, shape, colour, and location are processed, relative to the surrounding environment, in a so-called "scene-based frame of reference". In contrast, the dorsal stream which subserves visually-guided action codes information that is used for the control of bodily actions in an (effector-specific) egocentric frame of reference. Processing of information in this stream is taken to be unconscious.

Two sources of evidence ground the approach. First, there is a double dissociation of two neurological diseases, optic ataxia and visual form agnosia. Ataxic patients are unable to guide their movements based on what they see, although they are able to describe exactly what they are seeing. Visual form agnosia, in contrast, shows the opposite pattern: patients are able to fluently interact with seen objects while being unable to report what they see. Second, experiments on neurologically healthy subjects show that visual illusions do not affect visually-guided actions despite their strong effects on visual perception. These observations suggest that different neural processing streams underlie visually-based action, and visual perception, respectively.

2V and action as a condition on perception

Based on these central claims of 2V, Clark (2001) has argued against views that take action as a constitutive condition on perception (henceforth 'constitutive views'). Such an account is defended *inter alia* by Evans (1982), Hurley (1998), Noë (2004), and Gallagher (2005). Clark's argument goes as follows.

The first step is the observation that 2V questions an intuitively plausible picture about the connection between perception and action, what Clark terms the assumption of Experience-Based Control (henceforth EBC):

> Conscious visual experience presents the world to the subject in a richly textured way, a way that presents fine detail [...] and that is, in virtue of this richness, especially apt for, and typically utilized in, the control and guidance of fine-tuned, real-world activity (Clark 2001, p. 496).

This picture cannot be maintained in light of 2V. On 2V, it is not conscious visual experience that guides our bodily actions toward objects but the non-consciously operating dorsal system that links visual input to motor output. EBC thus has to be rejected.

As the second step, Clark claims that there is an implication between theories that posit a constitutive link between perception and action and EBC (p. 499): action can only be thought to be a constitutive condition on conscious perception if conscious experience is involved in action control.

To sum up, Clark's argument against the constitutive connection between action and perception is: 2V shows that EBC is wrong; EBC is an essential part of theories that posit a constitutive connection between action and perception; therefore, the view that perception constitutively depends on action cannot be sustained.

Is this argument valid? Briscoe (2009) objects to Clark's argument. Briscoe aims to show that Clark's EBC is not implicated in all constitutive views. His interpretation of EBC (call it 'EBC^B') is critical here. According to Briscoe, EBC^B consists in the claim that "visual awareness is *typically* implicated in visually based actions"[145] (p. 429). EBC^B is a claim about visually-guided actions and specifically about the way they are controlled, namely that typical instances of this kind of action are controlled by conscious visual experience.

It is not hard to see that a refutation of EBC^B does not conflict with the constitutive claim. Suppose EBC^B is false, i.e. central cases of visually-guided action are not controlled by visual experience but, e.g., by unconscious processing of the visual information. Does this speak against the constitutive view? No: claiming that central instances of visually-based action are controlled by other means than visual experience has no implications about whether visual experience is linked to visually-based action. The constitutive view implies that "visually-based action is implicated in all visual awareness of space" (p. 429), a claim which is not threatened by evidence to the effect that visually-guided actions are typically not controlled by visual experience.

Briscoe is not wrong in a way. It does not follow from any version of the constitutive view that every visually-based action is controlled by conscious visual experience, wherefore evidence to that effect does not affect these views. In the way he interprets EBC, viz. essentially as a claim about visually-guided actions, it is not implicated in constitutive views. The question, however, is whether Clark has a different understanding of EBC. It seems that he did not want his claim to be about visually based actions and what controls them but about visual experience and its functions instead.

According to this interpretation, the core idea of EBC would be that visual experience is a possible means to control bodily actions: "conscious visual experience [...] is especially apt for [...] the control and guidance of fine-tuned, real-world activity" (Clark 2001, p. 496). Following this, the focus of EBC would not be on the way visually-based actions are controlled but on the

145 The whole sentence reads: "And, hence, in contrast with EBC, ACT [action-oriented coding theory; a version of the constitutive view] does not claim that visual awareness of space is *typically* implicated in visually based action" (his emphasis).

question of what perceptual experience might be used for. And this would better fit the argumentative place that Clark envisages for EBC: namely, that it is implicated in theories that explain the contents of perception via direct links to action. For this to be possible, perception must indeed be so linked to action, otherwise the view would be suspect: if it was not possible to control one's actions on the basis of visual experience, how could one claim that the contents of visual experience constitutively depend on links to action? Accordingly, since 2V does challenge *this* direct connection between conscious visual experience and action, it is also a potential threat to constitutive views.

I think one cannot dismiss Clark's reasoning as easy as Briscoe thinks one can, i.e. EBC^B is mistaken. If one argues that the contents of perceptual space are constituted by links to action, as the constitutive views does, and if there is evidence that there is actually no such connection of perception to action, one has a problem. Hence, in general, I think that 2V has the potential to threaten constitutive views.

Let me turn to another argument that tries to establish a role for action to shape conscious experience in a specific way. Matthen (2005) argues, contrary to 2V, that processing in the vision-for-action stream grounds a certain feature of conscious visual experience, namely the feeling of presence of perceived objects. Thus, representing an object in a way that can be used immediately in acting on the object grounds an important feature of conscious visual experience. He uses the contrast of perceiving objects in the vicinity and perceiving objects in a depicted scene. He argues that depicted objects have no location relative to oneself, but are in a different space as it were. Thus, they are not present in the way that other perceived objects are present. He thinks this phenomenological difference is accounted for by representations of the dorsal system which are taken to be coding spatial information in an egocentric frame of reference, where spatial relations between subject and seen objects are specified relative to the subject, thus enabling the subject to act on these objects. He claims that dorsal-stream representations are only in play in perceptions of the environment, whereas there are no such representations of depicted objects.

However, this argument does not only conflict with what he argues elsewhere about dorsal stream processing – namely that it remains unconscious tout court. Furthermore, there are other possible explanations why we perceive depicted objects not as present in the way we perceive perceived objects as present. I think the most plausible answer is that we simply know when we are looking at a picture and when not. But this knowledge seems not based on the presence or absence of dorsal-stream representation: you can be fooled by hyper-realistic depictions of scenes, as in Trompe-l'œil paintings, to think that you are actually part of the depicted scene, as it were – and only when you realize that it is just a painting would your phenomenology change. Matthen would have to take one of two options here: One option is to argue that in perceiving Trompe-l'œil paintings dorsal-stream representations referring to the depicted objects are in play. But then the dorsal-stream representations cannot account for the difference in phenomenology once you have realized that you are only seeing a painting – since it is implausible that these simply vanish once you know that it is just a painting. The other option would be to argue that in cases of Trompe-l'œil pictures there are no dorsal-stream representations involved – but this would speak against his own account, given that the phenomenology in these cases is indistinguishable from perceptions of real environments.

Finally, I want to turn towards an argument against a particular account that supposes a constitutive link between action and perception, namely Noë's (2004) sensorimotor knowledge view. On his view, perception is constituted by practical knowledge about the sensory consequences that one's own movements would have. To see something to the right, e.g., is constituted by some practical knowledge of how the pattern of visual sensation would change were I to move toward the object. Brogaard (2011) argues that 2V challenges Noë's view:. "If the exercise of sensorimotor know-how is exhausted by action-guiding dorsal-stream representations, and Milner and Goodale are right that action-guiding dorsal-stream representations are unconscious, then perceptual experience is unconscious" (p. 1095). Following this quote, Brogaard presents evidence to the effect that dorsal-stream representations indeed are unconscious.

However, her argument is weak in multiple respects: she identifies sensorimotor knowledge with dorsal-stream representations without argument and she supposes that sensorimotor knowledge is realized in the cerebellum; however, both claims cannot be true: the cerebellum is not taken to be part of the dorsal stream at all. Furthermore, she does not take into account the difference between action-guiding representations and representations about sensory consequences of movements (which really are the basis of sensorimotor knowledge): it is not clear why the latter have to be processed in motor areas.

Arguments in favour of action control by conscious experience

Several attempts have been made in the literature to secure the action-guiding role of visual perception in spite of the evidence for the separation of two visual streams. The discussion has often centred on patient DF, a patient suffering from visual form agnosia. While she cannot report what she sees, she can quite normally interact with objects before her.

Wallhagen (2007) has argued that DF controls her acts by conscious visual perception but that she is merely not able to report what she sees because the contents of her visual experience are exclusively non-conceptual. However, a problem for this line of thought is that it is dubious why she also cannot report what she sees manually (e.g. reproducing the length of an object by thumb and index finger; cf. Whitwell et al. 2015) – something that might be considered possible based on non-conceptual representations alone. A more general worry might be that the very idea of bare non-conceptual but conscious representations is incoherent. For example, on Evans's (1982) view, where only those contents of perception can become conscious which *possibly* determine the contents of (some of) our thoughts, this possibility is excluded.

Mole (2010) has criticised Wallhagen's approach for failing to see what is really wrong with DF's perceptual contents. On Mole's account, DF cannot report what she sees because she has no conscious visual experience in passively viewing a scene due to her lack of "embodied demonstrative concepts". These would ground conscious experience of spatial features – but only when

DF actually acts on the seen objects. Thus, whereas in Wallhagen's analysis DF consciously perceives what is before her but cannot report her experience, on Mole's view she only perceives objects when she actually moves to act on them. Mole draws on evidence that DF is better in describing some object when she actually acts on it (Schenk and Milner 2006).

However, the problem with this approach (apart from the underdeveloped concept of an embodied demonstrative) is that the action-guiding role of conscious perception is supposed to be secured by an act that itself needs to be controlled in the first place. The actual performance of the act comes too late for grounding the content of the state that is supposed to control the movement (Wu 2013). Mole (2013) replies to that critique by arguing that the occurrence of the embodied demonstrative is not tied to an actual performance of the act. However, in this latter paper he only argues that the embodied demonstrative contributes to overall conscious experience, but not to *visual* awareness necessarily. This might be enough to get rid of the problem of unconscious action – it is not enough, however, to secure the action-guiding role of conscious visual perception.

Conclusion

I have considered various philosophical arguments that centre on the neuroscientific thesis that there are two functionally separated visual streams, subserving vision for action and vision for perception respectively. None of the arguments discussed are convincing. Briscoe's (2009) defence of views that take action as a constitutive condition against an argument presented by Clark (2001) rests on a misunderstanding of Clark's notion of Experience-based Control (EBC). Matthen's (2004) argument to the effect that dorsal-stream representations underlie the feeling of presence of perceived objects is unconvincing too. Brogaard's (2011) attempt to argue based on 2V against Noë's (2004) sensorimotor approach rests on a misunderstanding of the notion of sensorimotor knowledge. Wallhagen (2007) fails to establish that the visual agnosic patient DF actually controls her acts by conscious experience but is not able to report them. Finally, Mole's (2010; 2013) attempts to argue that DF actually consciously controls her acts by way of an embodied demonstrative also fail.

Let me note again that I do not think that 2V as it is formulated by its proponents is entirely appropriate. I only want to argue that these arguments taking into account 2V are mistaken and should be abandoned.

4 Egocentric Space as Self-Representation

4.1 Egocentric spatial representation as self-representation III: Argument from action as a constitutive condition on egocentric representation

In this final part of my dissertation, I want to argue that egocentric representation is to be regarded as a form of self-representation. Therefore, this chapter links back to the first part where I explicated the notion of self-representation and presented two arguments in favour of the self-representational nature of egocentric representation. The first of these was based on the supposedly relational content of egocentric representations; the second drew in a particular way on the design of the cognitive machinery. I argued against the basis of the first of these arguments by endorsing the idea of monadic spatial contents in egocentric representation (Ch. 1.2.2). I argued against the second argument that either it gets the constitutive basis wrong on which egocentric perception is based or, if 'cognitive machinery' is to be regarded in a broad sense, it is trivial and therefore lacks explanatory power. My aim in the rest of the dissertation was to develop a more substantial view about the constitutive conditions on egocentric representation, in particular spatial perception. In Parts 2 and 3, I discussed approaches that take action as a constitutive condition on spatial perception and presented various kinds of empirical evidence for the claim. Based on that evidence I ruled out several accounts and argued that those views that take the subject's capability to act as a constitutive condition for spatial perception are most convincing.

In this final chapter, I will first argue for a version of the capability approach that is inspired by Scheler's account. The conclusion of the discussion of the empirical evidence was that the most convincing accounts vis-à-vis the discussed cases were the ones developed by Evans and Scheler. I will present some considerations that favour a Schelerian view over Evans's. The most important task will be to spell out in detail what it means to have a grasp of one's own practical possibilities, and which of the practical possibilities are decisive that are open to one at any moment.

In the final step, I will argue based on this understanding of the constitutive link between spatial perception and action that egocentric representation is self-representational. Since the sense of what *the subject* can do (partly) constitutes perceptual space, the latter necessarily concerns the former. Therefore, even if the contents of egocentric spatial representation have to be regarded as monadic, as I argued before, egocentric representation is revealed as self-representational.

4.1.1 An argument for the Schelerian approach

The discussion in Ch. 2.1 through 3.2 revealed that there are many different accounts that propose a constitutive dependence of spatial perception on action (see Figure 6). The relevant notion of action that is proposed in the accounts varies substantively. I first ruled out accounts that rest on actual movements of either the subject or the environment by drawing on Hurley's contrast between instrumental and constitutive dependence. The core of mere instrumental dependence is that action is a means to enable the subject to perceive things it would not without movement. However, this does not mean that such movement is constitutive of perception. It is at most an enabling condition.

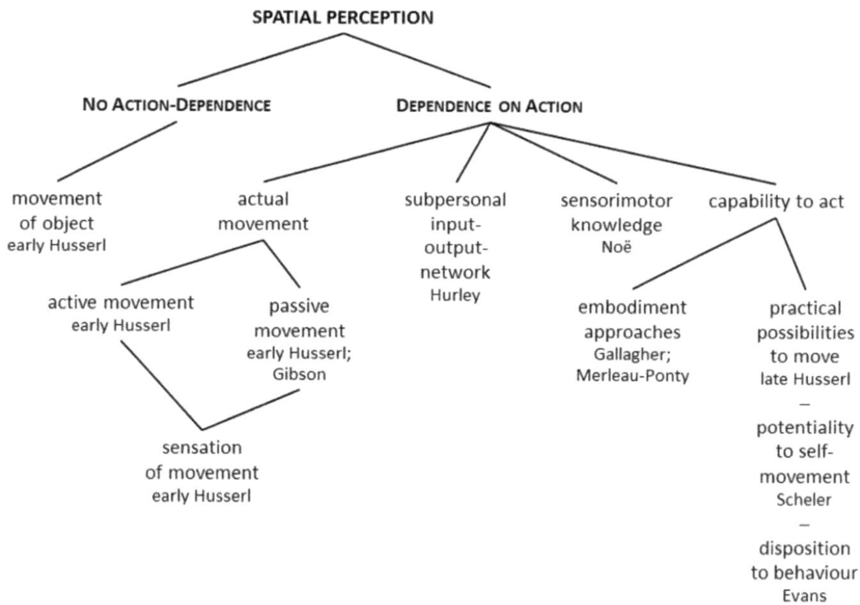

Figure 6: Accounts that have been discussed in Ch. 2.1 through 3.2.

After having dismissed accounts that rest on actual movements such as the early Husserlian account and the Gibsonian account, several approaches were still left. The empirical evidence discussed in Ch. 3.1 helped me to assess further the remaining approaches. I argued that the observations made by Proffitt et al. that indicate a role of output signals (such as motor simulation) count in favour of Hurley's two-level interdependence view. Her arguments are targeted against the general applicability of the classical Input-Output picture on which perception is exclusively constituted by processing of input signals. Against this picture, she puts forward cases where input patterns are constant but perception changes due to variations in output signals. However, Hurley's account only applies to particular instances of perception. She does not present a conclusive argument for the more general claim that the capability for spatial perception is dependent on the capability to act.

Concerning the sensorimotor approach endorsed by Noë, I argued that it is challenged by the results of inverted vision studies as well as studies on sensory substitution devices. Each report of these experiments includes an intermediate stage where subjects are able to interact with what they see without a concomitant re-inversion of perception or presence of spatial perception, respectively. I argued that these observations reveal that changes in the relevant knowledge about how sensation would alter with movement are not accompanied by changes in perception, contrary to what the approach implies. The sensorimotor approach relies on the idea that spatial perception is constituted by the subject's knowledge about how sensation would alter with movement. Therefore, it is falsified by evidence to the effect that substantial changes in that sort of knowledge have no effects on spatial perception.

Thus, we are left with the capability approaches. I discussed two accounts that focus on notions of embodiment. However, even if it could be made clearer what the partly metaphoric claims within these accounts (e.g. 'gearing') really refer to, and even if we could find a consistent interpretation of Gallagher's account, these approaches do not centre on action but rather on representations and functions of the body. Therefore, since they do not focus on the capability to act literally but on morphological properties of our bodies, I did not discuss them further vis-à-vis the empirical evidence.

The remaining possibilities are the capability views rooted in the works of Scheler and Evans, respectively. Both centre on the idea that dispositions to act are what partly constitutes spatial perception. The key difference between the two is that Evans holds that the presence of these dispositions is enough while Scheler argues that some grasp of them is essential. Both approaches seem to be in accord with the empirical evidence that I presented hitherto, hence it was not possible to decide between the two on that basis.

In this chapter, I will present some considerations that speak in favour of the Schelerian approach. First, though, it is necessary to get the approach in clear view. The crucial idea is the grasp of one's own possibilities that is implied in the account. However, Scheler's own writings contain no details about how the idea is to be spelled out. Hence, they can at best serve as a starting point in clarifying the idea. I will try to illuminate the idea *inter alia* by drawing on results from developmental psychology. I will argue that the relevant sense of one's capabilities is connected to planned action.

The second question that needs to be addressed concerns the kinds of actions the subject needs to have a sense of to enjoy spatial perception. I will argue that she needs to have a sense of her basic abilities to move toward objects, like turning, pointing, and reaching. This will provide a detailed understanding of the constitutive connection between this basic sense of capabilities and spatial perceptual experience.

Finally, I will come back to the question of why this approach is superior to the one that is inspired by Evans that rests on the mere presence of the dispositions to act. I will discuss Campbell's notion of causal indexicality and an experiment on full-body illusion to argue in favour of the Schelerian approach. My argument will rest on a consideration of whether causal indexicals correlate with the actual practical possibilities of the subject or with the sense the subject has of its own capabilities.

Accordingly, section 4.1.1 has three sections that each include one of the steps of the argument that I have just outlined. Let me start with the first point, which is to clarify the notion of a sense of one's own capabilities.

4.1.1.1 Sense of one's own capabilities[146]

The core of the notion of a sense of one's own capabilities seems to be that it provides the subject with information about the ways in which she can act. Thus, at least it is some informational state in that it conveys to the subject information about something (what this something exactly is will be discussed later). Beyond stating this core notion, various possible further specifications are *prima facie* possible – ranging from a very primitive sub-personal level state

146 In former sections on Scheler, I mainly used the term 'grasp'. Here, I switch to talk about a 'sense' of capabilities to indicate that the ideas presented in this section are independent from what he actually took the notion to be, which is not evident from his work.

to genuine knowledge had by the subject. I think that it is something in between.

Let me illuminate the idea by asking what purpose it serves. When do we need some understanding of what we are able to do in our daily lives? The idea that pops into mind is planning. In planning some action, one considers steps that need to be taken to reach a goal. Performing a planned, or intentional, action includes some appreciation of the instrumental structure that underlies these actions. It seems, necessarily, that I would need some appreciation of my abilities to be able to plan an action. How else could I determine the steps that I will take to reach my goal? Imagine you are in a familiar environment, let us say a small town, and you want to go to the supermarket. In figuring out a route to the supermarket, you will not only take into account known obstacles like houses, hills and rivers. Your planning will also respect in some way the things that you can and cannot do. You will not plan to jump over the stream but take the bridge – unless you are indeed able to jump over the small river. A bad jumper might take the bridge, a good one follows a route on which he would jump over the river. For the latter, the river is no obstacle. Thus, what you regard as an obstacle in your planning depends, in part, on what you take yourself to be able to do. Note that when you plan the route you usually will not consciously reflect on your capabilities. However, this does not mean that in planning you do not rely on a sense of what you can do.

Consider the contrary – planning an action without taking into account your capabilities. You would consider possibilities to reach some goal from an objective point of view as it were, abstracting away from the constraints that would shape your possibilities to move. By hypothesis, you would not take into account your own capabilities. However, this would imply in turn that you also could not take into account the features of the world you inhabit when planning. Since, whether something in the world is regarded as something one can act on or not seems to depend on the agent's sense of his capabilities (as was illuminated by the example of planning a route to the supermarket above). It follows that your action planning would not be constrained at all if you did not rely on a sense of what you can do.

The point about planning is to find an efficient way of coping with internal (related to your capabilities) and external (related to the environment) constraints to reach a goal. Planning loses its sense if there are no such constraints. If planning did neither rest on some appreciation of one's own practical possibilities, which would imply that there would not also be any external constraints perceivable, planning would be meaningless.[147]

Furthermore, *even if* you could make sense of the idea of planning without taking into account your capabilities to act in any way, you would end up with a mere consideration of abstract possibilities about how to reach your goal, but you still would not know what to do. Each possible alternative you planned would seem equally possible. It seems that at least a decision of which possibility can and should be taken needs some appreciation of your capabilities. Without this, your planning would not literally have led to a *plan* as it were.

In sum, I think that planning rests on some appreciation of one's own capabilities. The point of planning is that the subject figures out how to reach a goal and opts for one possibility. Without an appreciation of one's own capabilities, this is impossible.

It is important to note that planning in the sense I have discussed is not only involved in sophisticated actions such as walking to some distant place. I think it equally applies to basic kinds of action that the subject does intentionally.[148] Thus, even when you want to intentionally reach for something in your vicinity there is some planning involved since even for these simple acts you have numerous possibilities of how to execute them. For example, you want to reach for the sugar on the coffee table, which is at the opposite side of the table. The coffee can is in the middle. You have different alternatives of how to get the sugar, e.g. grasping either across or past the can. In a mini-

147 The point here rests on the way humans plan their actions. There might be creatures intelligible that plan actions and accordingly consider possible means without these means are constrained by their own capabilities but only by facts about the world they live in. Angels might be conceived of as such creatures.

148 By 'intentional' actions, I mean movements that are done at will by the agent and that cannot be exhaustively explained by reference to some reflex mechanisms.

mal sense, planning consists in deciding between these possibilities. The decision will in part depend on how you take yourself to be able to move in that situation. As before, if it did not depend on a sense of your capabilities, you would not regard the can in the middle as an obstacle, and you would have no grounds to choose one possible act over the other.[149]

One might ask just how basic this sense is, whether there are reasons to attribute it to young children or even animals. A clear case where it is not attributable is behaviour that can be explained by stimulus-response mechanisms. Such behaviour simply consists in executing some behavioural routine that is triggered by a stimulus. An explanation of such behaviour need not draw on any intentional states at all.

The question is what kind of behaviour *would be* indicative of a sense of the agent's own capabilities. In the ecological tradition following Gibson (1979), the idea that perceiving the environment is inseparable from perceiving oneself has led to the notion of the 'ecological self' as a primary form of self-knowledge that is already had by very young children (Neisser 1988; 1995; Rochat 1995; 2004). One of the elements that form this basic kind of self-knowledge is supposed to be an "awareness" of the subject's own action capabilities.

The following paragraphs explore the idea of this basic sense of own possibilities thus adding to the plausibility of a basically Schelerian version of the constitutive claim.

Evidence from developmental psychology

I want to discuss two kinds of studies that purport to show that young infants already possess a basic sense of their capabilities. The first focusses on reaching behaviour by 5–6-month old infants; the second focusses on moving down slopes by 8–14-months old infants.

In fact, the results of these two different kinds of studies could be used for two similar but slightly differing arguments along the lines presented

149 O'Brien (2007), pp. 163–8 argues in a similar way that intentional action implicates a grasp of the possible basic ways in which the subject can act. See below for some discussion of the idea of *basic* acts.

above. The first that rests on the observations by Rochat would be an inference to the presence of a sense of one's own capabilities based on the conditions on perceiving affordances. In particular, it would state that to perceive something as affording some action already implies that the subject has some sense of her own action capabilities in some way.

The second argument follows a different strategy. It does not rely on the conditions for perceiving affordances but on the conditions for planned action. The basic idea here is that the cases from Adolph et al. show that small infants can be said to plan their acts in a basic sense and this will include a sense of their own capabilities. I want to endorse this second argument eventually. In order to understand it better, however, it is helpful to start with a discussion of the first argument.

Developmental psychologist Philippe Rochat (1995; 2004) argues for the claim that the ecological self in Neisser's sense, which includes some awareness of what the organism is capable of, is shown by explorative, flexible and goal-directed behaviour. Thus, any organism that is capable of this kind of behaviour, i.e. behaviour that is purportedly not describable by mere stimulus-response links, has some sense of its action capabilities already (via having a sense of self in the ecological sense). Rochat's primary example is reaching behaviour. He suggests that children aged 5–6 months already rely on a sense of their own practical capabilities in reaching.

Let me shortly present the way he tested this hypothesis (Rochat et al. 1999). The basic idea of this study was to compare the reaching behaviour of infants that either can or cannot sit on their own. To be able to sit independently enables infants to reach further without falling over, thus extending their 'prehensile space'. The rationale of the study was to put two groups of children that differed only in their ability to sit in the exact same circumstances and to observe whether they differ in their reaching behaviour towards distant objects.

Infants were seated upright and strapped at the waist such that they were prevented from falling off the seat but were allowed to lean forward. They were presented with goal objects in four varying distances: the first distance was the nearest and above their toes. It functioned as a reference distance. The three other distances at which the goal objects were presented were each further away by 12 cm compared to the previous one. Objects at distances 1 (toes), and 2 (toes + 12 cm) were within prehensile space. Objects at distance 3 (toes + 24 cm) were at the limit of prehensile space so that they could only be touched by the infants with great effort. Objects at distance 4 (toes + 36 cm) or beyond were out of reach.

The relevant difference between the two groups of sitters and non-sitters was in reaching towards objects at distance 2.[150] Only half of the non-sitters reached for that object while more than 80 % of the sitters reached for it. The investigators ruled out that this difference is due to differences in size of the retinal image of the object. They recorded the varying distances between the forehead of the child and the object due to movements of the child and found no correlation between this distance and reaching behaviour. Thus, the initiation of reaching did not depend on the size of the object as it was projected on the retina.

Rochat et al. concluded that children at this age already have a sense of their own reaching capabilities that is in part determined by their ability to sit or not and determines the initiation of reaching toward distant objects. Thus, effectively, they explain the different patterns of reaching behaviour by a difference in the infants' sense of their capabilities. On their account, the "decision" (p. 137) made by the infants of whether to reach or not to reach is determined by whether the infant takes himself to be able to reach for the object or not. Thus, they claim that "reaching or not reaching by young infants is based on an accurate sense of their own body effectivity: a sense of their relative ability to maintain postural balance while reaching forward towards a distal object" (p. 138).

Thus, their basic idea is that a sense of one's own capabilities is a condition on perceiving affordances. Therefore, since the infants act depending on the affordances they perceive, it follows that they must have a sense of their capabilities.

150 The overall small number of contacts with objects at distance 3 did not allow for statistical comparison.

Prima facie, this seems a possible explanation of the results. However, one might wonder whether it is necessary to allude to a *sense* of the infants' capabilities to explain the results and not just to the infants' abilities *simpliciter*. I think that an answer depends on the particular description of these cases. The description that seems to underlie Rochat et al.'s explanation is that the infants perceive the goal and *decide* to either reach for it or not. This would imply some minimally intentional action performed by the infants and would make the case similar to my example above of reaching over the coffee table. Both cases include a representation of a goal object and a decision to reach for it. I argued above that this decision would involve some planning in the minimal sense of determining the way one reaches for the object and thus some sense of one's own capabilities to do so.

However, there is another possibility to describe Rochat's cases. The infants might perceive the object as reachable or not depending just on their actual abilities to sit. There is no need for a *sense* of their capabilities. To perceive that an object affords reaching might simply be dependent on the subject's actual abilities. It is possible that the actual abilities of the infants determine whether something is perceived as reachable or not without the infants' need to be aware of the fact that they are able to reach. Given this possibility, the sitters in Rochat et al.'s study just perceive the object as reachable and accordingly reach more often whereas the non-sitters do not perceive them as reachable and accordingly reach less often. However, this behaviour is not based on a sense of their own abilities but merely on their actual abilities that shape their affordance perception.[151] On this view, their behaviour is rather to be described as a reaction to what they perceive instead of a decision-based act.

Note, in addition, that Rochat et al. emphasize that the sense is "accurate". However, if they thereby mean that the infants' sense of their capabilities is necessarily correct, its explanatory value is dubious. Explaining in terms of *actual* abilities is as good as explaining in terms of a *necessarily correct* sense of abilities, if no independent support for the presence of the latter is provided.

151 Note that the authors themselves talk about perceiving affordances.

Rochat et al. do not provide such additional evidence for the presence of the infants' sense of their capabilities. Thus, mere reaching activities are not enough to ground the idea that infants have a sense of their capabilities.

In sum, the cases that Rochat cites in favour of his view are at best ambiguous. They do not unequivocally indicate that young infants possess a sense of their own reaching capabilities. This shows that the simplest form of reaching behaviour – to reach for some goal object presented directly in front in reaching space – is not indicative of a sense of one's possibilities. To be sure, the cases are compatible with the presence of such a sense – but they are not suitable to ground an attribution of it.

Part of the rejection of Rochat's argument thus is the rejection of the underlying premise that a condition on the perception of affordances is a sense of one's own capabilities to move. There is a less sophisticated way of explaining the perception of affordances in these cases, namely in terms of the actual capabilities of the subject.

Let me turn to the second argument that is based on studies from Adolph et al. that arguably include more than mere affordance perception and proper reaction to it. In the cases she focusses on, infants show more sorts of exploratory behaviour than mere looking and grasping. In particular, her cases suggest that infants try out different ways in which they could accomplish a task. This 'consideration' of various means to reach an end is the crucial difference to Rochat's grasping cases. This second argument for the presence of a sense of their own capabilities in young infants thus relies on explorative behaviour that is more complex and includes some way of planning process on the side of the infant and a decision, both of which is not necessary for explaining the reaching behaviour in the previous studies. In effect, this argument infers the presence of the basic sense of capabilities from the conditions on planning an action.[152]

Let me have a detailed look on Adolph et al.'s (1993; Adolph 1995) studies.

152 Neisser (1995) also suggests that these studies present a clear example of the presence of the ecological self, including a sense of their own capabilities. His proposal effectively is that the presence of the exploratory activities observed in these cases indicates the presence of a sense of their own capabilities. However, his statement is brief and he does not comment on the different kinds of exploratory activities that are observed in these cases.

These studies are not about infants' reaching behaviour but about their ascending or descending slopes of various degrees. Of particular interest is the way in which young infants descended the slopes. In particular, in contrast to ascending, where they tried to reach the top without hesitating, in descending, they first stopped at the edge and explored the situation. These exploratory acts consisted of looking from a distance, touching the slope with either feet or hands and probing different positions for going down (Adolph et al. 2000). Let me consider the details of Adolph et al.'s (1993; 1995) experimental paradigms and their observations.

Adolph et al. used slopes of varying degrees to assess the perception of affordances by young infants. Infants started on a flat platform, moved forward to the edge of the sloping middle section, went down / up, and finally reached another flat platform. A parent stood at the landing platform coaxing the infant to come. Adolph et al. (1993) had slopes of four different degrees of steepness (10°, 20°, 30°, and 40°), resulting in eight trials – one up and one down movement per degree. They compared the behaviour of two groups of infants. The first consisted of 23 14-month old *toddlers* that were all able to walk at least 10 feet freely. The second group consisted of 28 8–9-month old *crawlers* that used different means of locomotion (crawling on the belly, on hands and knees, or on hands and feet). Importantly, infants had no prior experience with descending slopes or slides.

In all ascending trials, regardless of the steepness of the slope, infants simply tried to reach the landing platform without hesitating – even if they were not always successful. These trials are not suitable to assess whether infants possess as sense of their capabilities. As in the reaching experiments by Rochat et al. discussed above, the behaviour can simply be explained by the (occasionally wrong) perception of the slope as affording ascending. The infants' behaviour is not indicative of some planning whatsoever.

Observations of descending trials are remarkably different. Moving down a slope is a more difficult task than moving up due to biomechanical differences in the movements (for details cf. Adolph 1995). This fact is reflected in the infants' behaviour. In contrast to the ascending trials, in descending trials, all infants stopped at the edge, hesitated, and touched the slope with either hands or feet before either going down or avoiding descending. However, importantly, there are also differences in the exploratory behaviour of toddlers and crawlers.

First, whereas toddlers often changed the way they went down after their exploratory movements, crawlers' exploratory activities did not lead to any differences in behaviour. Crawlers always simply used there default mode of going down (although *they were* in fact capable of different ways of going down, too). In contrast, toddlers changed from walking down to sliding down after touching steeper slopes. Thus, the practical consequences of exploratory activities differed between the two groups.

The second important difference concerns the exploratory movements themselves. In contrast to crawlers, toddlers showed an additional kind of exploratory activity, namely testing out different sliding positions. This was indicated by shifts in position that infants made before actually going down. The absolute numbers of shifts ranged from zero to eight, the overall mean was between one and two shifts of position. Infants only shifted position when they eventually slid down or avoided descending, but they did not shift positions when they eventually walked down. The majority of the infants figured out a safe sliding position and descended the slope.

Adolph (1995) replicated these results about toddlers' probing various positions before descending based on a higher number of trials. Furthermore, she used finer-grained and subject-relative measures of slopes, which enabled her to determine the 'walking boundary' for every infant. The walking boundary for each toddler was the steepest slope that it managed to walk down in two out of three times while failing or refusing to walk down a 2-degree steeper slope.

Her observations showed that toddlers probed different positions only before going down slopes that are at or beyond their walking boundary. The positions included "standing, squatting, kneeling, sitting, crawling, lying prone with stomach down, backing or using a pole as a railing or banister" (p. 740) and had to be maintained for at least half a second to be counted. Numbers of shifts increased with increasing degrees of slope and overall varied between zero and ten shifts. The mean exceeded one shift of position at slopes steeper than walking boundary. Interestingly, when infants probed more than one position, they often stood up and returned to their normal position before probing the next position.

Overall, the work of Adolph and colleagues indicates the presence of exploratory activities before moving down steep slopes starting at the age of eight months. It also reveals that there are at least three kinds of exploratory acts that need to be distinguished: exploration from a distance by vision, exploration by direct contact with the slope by hands or feet, and exploration of alternative means of reaching the goal (Adolph et al. 2000). Exploration of alternative means is only observed in the behaviour of infants that are capable of walking already (toddlers) but not in those that are not yet able to walk (crawlers). Furthermore, the behaviour of the younger infants is apparently not influenced by their exploratory movements. Thus, they did not choose means of going down the slope other than their normal way of moving. In contrast, toddlers' way of going down the slope is more flexible and attuned to the situations they face.

Let me come back to the initial question of whether exploration activities are a clear sign of a sense of one's own capabilities. I think the claim is not true if put in this general way. On my view, the first two forms of exploratory behaviour are not enough to ground such an attribution. Exploration by vision is not different from what the reaching infants in Rochat et al.'s study did. This is simply a means of gathering sufficient information to determine the affordance presented by the object or situation. I think the same should be said about exploration by direct contact. This can also be regarded as a way of finding out what the environment affords – only that it draws on a different kind of information, namely tactual and proprioceptive instead of visual. I think that both of these kinds of exploratory behaviour are activities that enable the infant to perceive what the environment affords. However, it does not show that the infant has some sense of her own capabilities.

I think that only the presence of the third kind of exploratory movements that consists in the probing of alternative means to move down the slope indicates the presence of a sense of their own capabilities. Note that the presence of this kind of explorative behaviour is the crucial difference between the behaviours observed by Rochat, and Adolph, respectively. I think it is reasonable to suppose that what toddlers are doing in that situation is planning how to reach their goal. They 'consider' alternative means and eventually chose one. It is not the same kind of consideration that adults normally do, since it is done in an active way and not by merely thinking or imagining alternative means.[153] Nevertheless, it is a way of planning. Toddlers are aware of the risk of falling down when walking and in response try out various alternatives to go down. To do so, they need to have a sense of their alternatives, which they try out at will. They are not merely reacting to something that they perceive. The descriptions of these cases do not suggest that toddlers just arbitrarily change their position. Instead, they seem to take different positions at will. Support for this interpretation comes in part from the fact that they often returned to their initial positon in between trying alternative positions.

Overall, in a basic sense, they decide based on a planning process that involves probing of different means via taking various possible positions of going down the slope. Thus, their behaviour can be counted as intentional acts, i.e. actions that are done at will and based on a decision that is grounded in a planning process that takes into account alternative means of reaching a goal.

It is certainly not justified to think that the sense in question is a global sense of all of the infants' capabilities. Rather, one can only infer a task-specific sense of their own possible means to reach a particular goal. In addition, since the sense is present at this early age, it is independent of linguistic or conceptual representation. To explain these cases it is not necessary to assume that the children draw inferences based on planning. Furthermore, the sense need not be accurate at that stage. The probing of the different possible alternatives itself might in turn help to refine the sense of what is possible.

153 However, when performing complex tasks in new situations, adults also sometimes probe different ways by actually moving.

I take it that I have illuminated the idea of possessing a basic sense of one's own capabilities sufficiently. The evidence discussed implies that it is present in 14-month old infants already. It is indicated by intentional behaviour that draws on some planning process, which takes into account alternative means to reach the goal resulting in a decision for one of the possible alternatives.

Now that I reached an understanding of the sense of one's own capabilities, I want to turn to the question of which *kinds of actions* are crucial for spatial perception. This is the second step of the argument for the Schelerian approach that I want to defend.

4.1.1.2 Relevant capabilities are basic movements

What kinds of actions must the subject have a sense of to be capable of spatial perception? Intuitively, it is unlikely that it is dependent on sophisticated action. It would be strange to suppose that one must have a sense of one's ability e.g. to tie one's shoelaces to be able to perceive them. I think what essentially matters is the basic repertoire of bodily movements like turning, approaching, pointing, and reaching that one can execute with respect to what one sees. It is plausible to restrict the range of intended actions in that way since it is plausible to think that e.g. children with a limited behavioural repertoire perceive spatially. One of the core ideas of the approach is that the spatiality of perception is grounded in the sense of one's capabilities to *move* with respect to what one sees.[154] Put differently, the location at which you perceive an object is determined by the movements you would execute with regard to the object if you had an intention to do so. Thus, the movements that are sensed as possible constitute the spatial contents of perception.

On what basis can we argue for a specific set of movements as crucial? First, one might consider again the empirical evidence that speaks in favour of the constitutive view. The study on Sensory Substitution Devices (SSDs) reported by Siegle and Warren (2010) indicates that the ability to point at an

154 Scheler (1973), p. 333 also speaks of the sense of one's possibilities to move ('Bewegungsmöglichkeit'; Scheler 1927, p. 312).

object is enough to perceive it at a distinctive direction and distance in egocentric space.

In their study about distal attribution by subjects wearing an SSD (cf. Ch. 3.1.4 above), the experimenters used a tactile SSD that converted signals gathered by a small video camera that was mounted on the subject's finger into tactile stimulations on the subject's back. Blindfolded subjects had to move their arm and finger; they were allowed to bend the arm only at the elbow and the shoulder. When the camera recorded the target object, they received a tactile stimulus on their back. The task was to estimate the distance of the target object in three sessions with 48 hours between each session. The results indicated that one group of participants, namely those that should base their estimation on an intuitive sense of the distance of the object but should not try to figure out the distance by triangulation based on the posture of their arm, perceived the object at a determinate location in space before them. Above, I discussed whether the evidence is sufficient for attributing spatial perception. Let us assume that it is enough, and the participants really perceived the object at a particular location before them. This implies that they perceived the object at a certain direction and distance relative to themselves. This would suggest that, at least for the case of near space (the object was located somewhere between 40 cm and 130 cm away from the subject), the ability to point toward the object – *in alternate ways* – is sufficient to perceive its direction *and* distance in egocentric space.

Beyond this, it is hard to find appropriate evidence from these empirical studies reviewed above about which kind of behaviour is crucial. For example, in the only study on inverted vision that systematically tested for re-inversion of the visual field, Sekiyama et al. (2012) report tasks that indicated a re-inversion of the visual field. These tasks consisted in manual or verbal reports about the location (either left or right) on which participants perceived a visual, or somatosensory stimulus, respectively. However, these tasks were designed to test for how the visual field is oriented but they did not test for the kind of behaviour the subject needs to be capable of for the visual field to re-invert.

I think it is no coincidence that Evans and others relied on thought experiments in figuring out what the subject must be capable of doing to be able to perceive spatially. As already discussed, Evans argued that a mere difference in behaviour in response to a difference in spatial location of the stimulus is not enough. For example, a subject that would respond to a sound coming from ahead by raising the left arm and respond to a sound that comes from behind by raising the right arm need not perceive these locations as such. It might just be sensitive of some difference in the sounds that co-varies with the difference in spatial location.

Thus, what is needed is an action that is constrained by the *location* of the stimulus. I think what Evans, Scheler and others have in mind is some sort of *directed* spatial movement. However, they did not elaborate on the details of the movements that are necessary. Let me try to spell out this claim by differentiating between different dimensions of spatial perception like direction and distance, or near space and far space.

To be able to perceive something at some direction, e.g. a tone as coming from the left would mean to have a sense of one's ability to turn toward the sound source (thus, not in a stimulus-response manner but as part of a planned action). A movement of the head toward the sound seems enough. Turning seems to be the basic kind of movement that is constrained by direction. You would not believe someone who told you that he could perceive a sound as coming from the left unless he was able to turn toward (or away) from it systematically. It seems that turning-towards is sufficient for perception of direction in near and far space. In a minimal sense, turning can refer to head movements or even eye movements. This would suggest that the basic ways in which one can shift one's attention at will that include bodily reactions are enough to ground the perception of direction.[155]

Perception of distance seems to require more than mere turning towards the object or a sound source. For near space, pointing in different ways seems

155 Since I rely on the idea that basic sorts of *movement* are decisive I do not consider covert shifts of attention, i.e. shifts of attention that do not include any overt bodily movement. An example for these shifts would be looking at a picture and attending to different features of it without moving the eyes or other body parts.

sufficient to perceive objects at a particular distance as is indicated by Siegle and Warren's study on SSDs discussed above. Otherwise, distance perception might be constituted by some sense of what it would take to approach the object. Objects that are very far away, like stars, can still be pointed at or turned towards, but the perception of them does not include a distinctive distance. This might be explained by a lack of a sense of what it would take to approach these objects.

We are always capable of a plethora of movements towards perceived objects at a time. For example, hearing a sound as coming from some direction does not only allow for turning towards the sound but at the same time, we can e.g. move towards the sound in an attempt to approach it. I think one cannot dissociate these two kinds of movements with respect to a given sound; the ability of one entails the ability of the other, *ceteris paribus*. Thus, the decision to regard turning as the primary movement relies on two premises. First, the aim is to express the *minimally sufficient* behaviour that has to be sensed as possible for oneself to be capable of spatial perception. This premise follows from the notion of constitutive explanation that is not aimed at sufficient conditions but on the necessary conditions that are together sufficient to explain some phenomenon (cf. Ch. 2.1.2 above). In other words, the aim is to find the minimal set of conditions for the presence of spatial perception.

The second premise is that turning is a more basic behaviour than approaching. This claim can be warranted by arguing that whereas approaching a sound is dependent on the ability to turn towards it, there is no such dependence in the other direction. Being able to turn does not depend on being able to approach. Thus, turning is more basic than approaching.

Talking about *basic* behaviour in this sense is similar but not identical to the notion of (teleologically) basic actions that is discussed in action theory (Hornsby 1980; Searle 1983; O'Brien 2007). Basic actions are defined as intentional actions one can do without doing something else intentionally. For most people, tying their shoelaces is a basic action. However, for children that are just learning how to do this it is not a basic action. They would tie their shoelaces by doing something else intentionally, e.g. moving their fingers *like this*. Thus, the difference between basic actions and non-basic actions in this context is not a question of complexity but of ability, or know-how.

In contrast, in arguing that the basic repertoire of movements is what is crucial for constituting spatial perception, I rely on a slightly different notion of basicness. One might put it in saying that these movements are the basic movements that can be basic actions. Thus, basic movements like turning, approaching and pointing are not only teleologically basic in the sense that they can be performed intentionally without doing something else intentionally – they are also basic in that they are among the most simple acts in terms of complexity and coordination demands that humans are capable of. This means that the abilities you bring to bear in these basic movements are part and parcel of the more complex and demanding actions you are capable of.

I am aware that this is not a proper argument for taking turning, pointing and approaching as the basic crucial movements. Nevertheless, I hope to have made at least my choice plausible and my reasons transparent for defending this view. I take it that the range of movements that is necessary is not completely determinable *a priori*. In addition, as the spatial perceptual abilities develop there will be different capabilities decisive. Arguably, expert spatial perception relies on a different set of capabilities than the basic cases of spatial perception that are present at young ages such as in the cases from developmental psychology I discussed above.

Sense of capabilities as motor simulation, or body schema?

I want to consider whether the sense of one's own capabilities as outlined can be identified with some other already developed ideas. One option would be to regard it as a form of motor simulation like imagined actions. These are supposed to include everything about a particular bodily action except for its execution. Hence, they also include a representation of the means necessary to reach some goal (Jeannerod 2001). Therefore, one could contend that the sense of one's own practical possibilities is a set of these motor simulations. However, it seems to me that motor simulation itself draws on an appreciation of what one can do. Imagined actions are about specific bodily movements that are not carried out. But imagining some bodily actions already includes some appreciation of what one can and cannot do. I think we are searching for an underlying state that equally informs imaginations of actions and actual actions.

A possible candidate for such a state might be the body schema. De Vignemont (2010) argues that the body schema underlies both actual action execution and motor imagery. Furthermore, on her view, the information embodied in the body schema is not only exploited by unconscious processes but it is accessible to consciousness in e.g. imagination of actions as well (pace Gallagher 2005). This would fit well with the observation above that the sense of capabilities is present in conscious planning as in the case of how to get to the supermarket, and with the supposedly automatic planning of descending movements by young children. Nevertheless, this proposal also seems to slightly miss the point. The body schema is a *body* representation that draws on different informational sources like proprioception and vision to represent the properties of the body that are used for action like size of limbs, angles of joints and current relative positions of body parts (de Vignemont 2010; Wong forthcoming). This means that it does not itself include any information about possible acts but that it includes information about the body for action control, planning and imagination. For us as embodied creatures, action capabilities depend in part on our body properties, but it seems false to suppose that the mere representation of these properties that are crucial for action already suffices for providing a sense of what one can do.

It seems that neither motor stimulations nor the body schema is the right kind of notion to capture the idea of a sense of one's capabilities. They might be related and play a role in the generation and updating of the latter but they cannot be taken themselves as such a sense. Therefore, I think it is not possible to identify the sense of one's capabilities with one of these notions.

An alternative proposal: the 'action space' model of conscious visual perception

Finally, let me turn to the only proposal of which I know that formulates a similar constitutive view. Ward et al. (2011) claim that visual experience "consists in practical knowledge of our own possibilities [...] for action" (p. 376). Thus, they endorse the basic idea that some sense of one's own practical possibilities is a constitutive condition on perception. How do they conceive of the relevant sense of capabilities? They regard it as "knowledge concerning the space of apparently-enabled intentional actions" (p. 389). This is still quite vague. How can it further be cashed out? First, the authors themselves admit that 'knowledge' is not the right notion since one can be wrong about one's capabilities. They rephrase it as "unreflectively taking herself to be poised over an action-space" (p. 390), thus getting closer to my own formulation of a 'sense of capabilities'. Second, they admit that the state is not dependent on the possession of concepts or language as I did above, too. Third, it is clear from their various formulations that they refer to a sense of one's *current* capabilities relative to what is perceived in a particular situation. Thus, they focus on the possibilities one takes oneself to have with regard to a specific context. Furthermore, this sense of possibilities needs to be integrated with overarching plans on their view.

The last two points are of utmost importance since they provide the details of the constitutive dependence between perception and action. I think Ward et al. are on the right track but I am not sure whether I would completely agree with them. The reason is that they provide several formulations of how they conceive of the constitutive dependence that together form an ambiguous picture. Their formulations include "direct knowledge of space of actions", "appreciation of action opportunities", "planning for action", "second order disposition", or the "ability to adjust actions and intentions in the light of encountered affordances" (pp. 383–4). Maybe it is possible to link these ideas to the case of the descending infants and their planning that I discussed above.

A problem for figuring out what their explanation of spatial perception is consists in their wide scope. Ward et al. argue for a constitutive dependence on action of all sorts of contents and also of the phenomenology of perceptual experience. That is why much of their paper is focussed on the idea of epistemic actions like grouping, categorizing, sifting and sorting and how this relates to perception of shape and colour. When it comes to the actions that are part of the supposed 'action space' that grounds spatial perception, they only state that these actions are to be described at a rather coarse-grained level that does not include fine-grained details about the properties of the bodily movements. The examples they provide are raising a glass in different ways (p. 383, n. 9). On my view, however, it is not necessary to refer to such sophisticated behaviour.

Overall, Ward et al.'s account about spatial perception seems to be underspecified in two regards. First, they do not provide any details about the actions they take to be crucial for constituting perceptual space. As I argued above, it is unlikely that sophisticated kinds of actions are needed to be able to perceive spatially. However, Ward et al. do not present any considerations about the kinds of actions necessary for spatial experience. This results in an unsatisfactory picture about how spatial perception is constituted in detail. I argued that it is constituted by the basic actions that are constrained by the spatial dimensions of perceptual space. In contrast, Ward et al. do not provide any consideration about the kinds, or aspects of action that constitute spatial perception.

In addition, despite their various claims to the end that their notion is supposed to be quite basic and independent of conceptual or linguistic representation, the details of their notion of a knowledge of one's action space are not clear. I think that in discussing the behaviour of toddlers, I provided a more substantial notion of the relevant sense than they did in their paper.

The lack of detail in describing both the nature of the sense and the details of the movements that the subject needs to have a sense of to be capable of spatial perception makes it difficult to compare the approach with mine in detail. I think the core of both accounts is quite the same in taking intentional, or planned action as the relevant notion of action (even though I understand my claims to be exclusively about spatial perception). I think that if they would develop an account about spatial perception in detail based on their general framework it would be very similar to my own. However, the wide scope of their account and the accordingly high level of generality on which they develop their account in that paper prevent a final judgement about this question.

To sum up, the position that I want to endorse is that spatial perception constitutively depends on a sense of one's own basic capabilities to move with regard to what one sees. In drawing on a sense of the possibilities, instead of their mere presence, and in supposing that a kind of planned behaviour is crucial, this position is different from the one that is defended by Evans.

Let me finally consider two lines of thought that support my view instead of Evans's approach. This is the last step of the argument toward the Schelerian approach that I want to defend.

4.1.1.3 Arguments in favour of the view

As a first way of arguing in favour of the view that a basic sense of one's abilities is crucial, I want to consider the notion of causal indexicality as it is proposed by Campbell (1994), pp. 42–6. An example for a causally indexical property that might be applied to objects is 'within reach'. What is the basis of applying such a property to some perceived object? Certainly, it has to do with what the subject is capable of doing. To represent something as within reach, or as too heavy, or as too hot, depends on one's range, strength, or heat tolerance, respectively. The significance of these representations lies in their immediate connection to action. As Campbell writes, "[the] most immediate effect of judgements made using this notion [i.e. being within reach; F.P.] is that the subject will try to contact things within reach but will not try to contact things judged to be out of reach" (p. 44). Causal indexicals thus refer to (features of) objects *and* directly shape the ways the subject would interact with the object if he wanted to. In other words, via causal indexicals one sees objects as having certain properties (that depend, at least in part, on the subject's powers) and these very properties are not only recognised as such in a distanced way but they are at the same time responsible for the way we deal with so represented objects.

Now the question is in what way these causal indexicals depend on the subjects' powers. Is it really reasonable to propose that they are indexed to the powers of the representing subject *simpliciter*? Alternatively, is it rather the case that they are indexed to some basic sense the subject has of its own powers? I think the latter is the case. This makes intelligible certain situations that would not be intelligible otherwise. For example, imagine you are Neil Armstrong, having just landed on the moon. Buzz Aldrin asks you to lift up some piece of rock since he wants to take a heroic picture of you. Since the rock is quite big, you represent it as too heavy, although, in fact, you are perfectly able to lift it up due to the decreased gravitational force on the moon. Your representing the rock as being too heavy was clearly dependent on some sense of your powers, not on your actual capabilities. Since you did not attune to the circumstances on the surface of the moon, you misperceived your abilities. Another example would be a case of phantom limb where someone represents something as within reach, while he is in fact unable to reach it. These cases describe situations in which one's actual capabilities and one's sense of one's capabilities come apart. Intuitively, it is plausible to suppose that the perception of causal indexical properties is tied to one's sense of and not to one's factual abilities. If the causal indexical predicates are simply dependent on our actual capabilities, these situations would seem to be impossible.

A way to test this claim experimentally is to use virtual reality (VR). Van der Hoort et al. (2011) use VR to induce in participants the impression that they own a radically different body – which is either much bigger or smaller than their actual body (cf. Piryankova et al. 2014). They tested whether distance and size estimations varied due to the illusion. They found significant effects on verbal and manual estimations of size and distance depending on the size of the body that participants take themselves to own, thanks to the illusion. Based on these findings the authors claim that "the sense of one's own body [size] affects how we visually experience the world" (van der Hoort et al. 2011, p. 7).

In light of the results, this claim certainly sounds right. However, taken literally, the claim would present an objection to the view I want to defend. On that account, the sense of one's capabilities is not crucial, rather a sense of the size of one's body. However, as I have said before, it is plausible to assume a close connection between a representation of one's bodily parameters and the sense of one's capabilities since the latter are to a certain extent constrained by the former. Hence, it is plausible to claim that there are systematic connections between representing relevant bodily properties and one's sense of what one can do as outlined above. Given this systematic correlation, the results of van der Hoort et al. are ambiguous and may be explained by two hypotheses. One is their own, which draws on the represented size of the body. An alternative hypothesis is that a sense of one's capabilities is what is responsible for the shifts in perception of size and distance. Both hypotheses seem equally plausible given the results (cf. Jackson 2014).

Given that we can explain the results by drawing on a difference in the sense of one's own capabilities, this study can help to argue in favour of the Schelerian approach instead of Evans's. They present a case where one's factual abilities do not change but only one's sense of one's abilities. Since there are systematic changes in spatial perception resulting from the illusion, it does not seem to be the case that one's actual abilities (which are not affected by the illusion) are constitutive of spatial perception, but one's altered sense of one's own abilities is.

I have presented two lines of thought that speak in favour of the view that a sense of one's capabilities is a constitutive condition on spatial perception. The first was based on Campbell's idea of causal indexicality that arguably relies on a sense of what one can do instead of relying on one's actual capabilities. This claim is underpinned by thinking about what would happen if actual capabilities and the sense of capabilities comes apart as in the case of the astronauts on the moon. The second reason for endorsing the 'sense of capabilities approach' relied on the same consideration, but this time it is not a mere thought experiment but one centred on using VR to alter the sense of capabilities by inducing the illusion that one possesses a small- or large-sized body. Verbal and manual estimations of distance and size vary systematically with the presence or absence of the illusion, thus suggesting that spatial perception depends constitutively on a sense of one's own action capabilities – and not on one's actual capabilities that are constant during the experiment.

4.1.2 Summary of the account

Let me sum up the position that I have reached based on the discussion of the preceding chapters. On my view, spatial perceptual space is constituted in part by the subject's sense of her own basic capabilities to move. This account is inspired by Scheler's approach in that it relies on a sense of one's practical abilities. However, Scheler did not spell out many details beyond this basic idea. I argued based on insights from developmental psychology that the relevant sense is present as soon as the subject is capable of planned action, which is based on some decision that takes into account alternative means of reaching a goal. Deciding which organisms are capable of such behaviour is an empirical matter. The evidence I presented indicates that it is present in 14-month old infants.

I argued that in specifying the kinds of movements one needs a sense of, one should propose the most basic movements which can be described as being constrained by the spatial dimensions of perceptual space. I argued that turning-towards is the basic movement that is related to the dimension of direction and that pointing-towards and approaching are the basic movements that are related to the dimension of distance (for near space, and far space, respectively).

My position is based on a variety of empirical evidence. The most important sources of evidence are studies on inverted vision and on sensory substitution devices. However, the account is also shaped by seemingly conflicting evidence that is part of the dual visual systems approach. The discussion of each kind of empirical evidence pointed toward the idea that the mere presence of dispositions to interact with what is perceived is not enough and that an account that draws on planned action and a sense of one's own capabilities fares better in accommodating the results. However, it was always possible to consider ways in which an Evansian picture might be developed that could equally explain the results. I take it that the considerations I presented in this chapter with regard to the notion of a basic sense of one's own capabilities, the idea of perceiving causally indexical properties and the results from the full-body illusion studies, add substantial plausibility to endorsing my view instead of Evans's. Overall, my position presents a well-motivated alternative to the predominant sensorimotor approach and Evansian approaches to spatial perception that fares best in accommodating relevant empirical evidence. Thus, spatial perception does not consist in some knowledge about the sensory consequences of one's own movements, and it is not constituted by the dispositions that are linked to incoming information. Instead, the way one perceives the world spatially is constituted by a sense of one's own capabilities for movement that enables one to plan one's actions and act intentionally.

4.1.3 Egocentric spatial representation as self-representation

Finally, let me come back to the initial question of this dissertation. The question is whether egocentric representation, as instantiated primarily in egocentric perception, is a kind of self-representation. In elaborating on the notion of self-representation in Part 1, I argued that self-representation as a kind of representation is not dependent on the *content* of these representations but on the functional role they have. In particular, I argued based on the works of Perry that what is distinctive of self-representation is that they concern the representing subject. However, it is neither necessary nor sufficient for self-representation that the subject represents itself. Instead, what marks the notion of self-representation is the immediate way in which instances are relevant for the actions of the subject. I also argued that this immediate relevance for action can be accounted for by drawing on the constitutive conditions on which the representation is based. A first argument, the argument from design as I termed it, proved to be unsuccessful since it relied on an inadequate understanding of the constitutive conditions. Now that we have a clear understanding of the constitutive conditions on egocentric representation, it follows from these that egocentric representation is a kind of self-representation.

It is a natural consequence of the account of spatial perception that I have defended that egocentric representation is a form of self-representation. Thus, egocentric space is self-representational since it constitutively depends on the subject's sense of *his own* capabilities to move. If the arguments in the preceding chapters are right, the way in which one spatially perceives the world is determined by a sense of one's own capabilities for moving. Thus, in a way, the spatial outline of perception is a correlate of this basic sense of one's own capabilities to move. I can think of no better way for securing the immediate relevance for action of egocentric representation than this. If the ways in which I represent the vicinity are structured according to how I take myself to be able to move, it seems obvious that I can directly use these representations to execute my actions. There is no need to figure out how to approach some object, point at it or grasp it since this is already implicit in the way I represent the world. The world as it is presented in egocentric perception is always already shaped by the subject's sense of his basic movements and thus can be acted upon in an immediate way.

In my initial discussion of egocentric representation in Ch. 1.2.1, I spelled out the immediate relevance for action that egocentric representations have in contrast to allocentric spatial representation that is also relevant for acting – but not in a similar immediate way. The idea was that whereas you might draw on allocentric spatial representations in finding your way, you need to have an egocentric representation of at least one of the places of which you might know how it is related to other places to act in a reasonable way. Thus, allocentric spatial representation is only useful for acting when it is translated into an egocentric representation that specifies the relation of the places with regard to the subject. The necessity of this translation process marks the border between 'immediate' relevance for action and indirect relevance for action. Representing an object or place egocentrically allows for acting on it without further translation in contrast to allocentric representation. My claim is that this property of egocentric representations is grounded in their constitutive dependence on the subject's sense of his own basic capabilities to move. This means that you immediately know how to interact in basic ways with egocentrically represented objects since without this knowledge you would not represent them in the first place.

4.2 Conclusion

In this dissertation, I have defended the claim that egocentric representation is a kind of self-representation. After introducing the idea of self-representation, I considered three arguments in favour of that main claim. The first was based on a relationist view about the contents of egocentric representation, which states that the representing subject is explicitly represented in the contents of egocentric spatial representations. I argued against this view mainly by alluding to work of John Perry's who defends a monadic view about the spatial contents of egocentric representation. The second argument to the effect that egocentric representation is to be taken as self-representational I considered was based on claims about the design of the 'cognitive machinery' that subserves egocentric representation. I argued that this approach is inadequate since it gets the constitutive conditions of egocentric representation wrong. The third, and to my mind superior, way of arguing for the main thesis can be considered as a way to compensate for this shortcoming of the argument from design. By arguing that action is a constitutive condition on egocentric representation, one does not only get the constitutive conditions of egocentric representation right but it also opens up the possibility for a well-grounded argument for the main thesis. Therefore, this step of my argumentation received most space.

In the course of developing my position I first made clear what is meant by 'constitutive condition' in this context. Then, I presented some of the historical background of this way of explaining spatial representation. In Ch. 2.2 and 2.3, I provided a systematic overview of different approaches that are defended to date, made clear their differences and discussed issues that are connected with the proposals. The aim here was to present the possible options to argue for the constitutive thesis and to set the stage for developing my own view. In Part 3, I turned towards empirical evidence. I presented three sources of such evidence that speaks in general in favour of the view that action is a constitutive condition on spatial representation. On closer examination, it was revealed that the evidence is not as unambiguous as it appears at first sight. All three kinds of evidence face some difficulties with respect to the design of experiments, interpretation of subjective reports, or significance of experimental results. However, I concluded that in sum there are good empirical based reasons to take the capability to act as a constitutive condition on spatial representation.

This part also helped to reveal problems of one of the most prominent approaches in the literature, namely the sensorimotor approach. The discussion of the empirical evidence thus helped to sharpen the view about what it means to give constitutive conditions on the capability for spatial representation, it presented reasons to hold that the capability to act is such a constitutive condition and it revealed that certain particular approaches are not suitable to argue for the main thesis. All this highlights the importance of empirical investigation even for such basic questions as about the constitutive conditions of spatial representation. Therefore, it was also reasonable to look at empirical evidence that would seem to speak against the constitutive dependence of spatial representation on action. I considered one well developed and widely acknowledged account of the processing of visual information, the dual visual systems approach that proposes only a somewhat indirect relationship between both capacities. However, in discussing the details and empirical bases of the account I argued that it is no empirical threat to the constitutive claim.

I argued that two views fare best in accommodating the empirical results. This were the two capability approaches rooted in the works of Evans and Scheler, respectively. The difference between these approaches lies in the supposition of a *sense* of one's own capabilities that is part of Scheler's account in contrast to Evans's. On Evans's view, the mere presence of behavioural dispositions that are linked to perceptual input is enough to constitute the spatiality of perceptual representation. In the final chapter, I have put forward considerations that speak in favour of a view that is inspired by Scheler in that it posits a sense of the subject's own capabilities for basic movements as a constitutive condition on spatial perception. To spell out the details of this account I relied on results from developmental psychology that indicate that 14-month old infants already possess the relevant sense of their own capabilities. Finally, I argued based on Campbell's notion of causal indexicality and studies on full-body illusions in favour of this approach and against the Evansian view.

The final step of the overall argumentation was to acknowledge that it is a natural consequence from this theory of spatial representation that it is a kind of self-representation. Since egocentric space is constituted partly by a sense of one's own practical possibilities its self-representational character that consists in its immediate relevance for action is revealed.

Overall, my dissertation connects claims about spatial perception (as the paradigmatic case of egocentric representation), action, and self-representation. If I am right, these three basic capabilities of humans and other species are deeply intertwined. The capability to spatial perception cannot be understood without taking into account the practical capabilities of the subject, something that has been expressed by the claim that "only agents are perceivers" (Schellenberg 2007, p. 603). However, although being true to my mind, this claim invites many further questions. Does the constitutive connection between action and spatial representation hold at the level of particular instances or at the level of capabilities? Are claims about the embodied nature of representing organisms inevitable? What exactly is it that is linked to action and explanatory of the constitutive relationship: knowledge or expectations about consequences of actions, sub-personal motor mechanisms, the largely automatic and pre-reflective workings of our world-attuned bodies, the capabilities we have to act on what we see, or some basic sense of our practical possibilities? I hope to have given some well-motivated answers to these questions.

The explication of egocentric spatial representation as self-representation can provide a starting point toward approaching one of the deepest and intriguing themes in philosophy: self-consciousness. One plausible approach is to start by explaining more basic and simple kinds of representation that at least share some features with self-conscious states. One such feature certainly is the special functional role that these states have and that is well brought out by examples such as Perry's shopper and Mach's shabby pedagogue presented at the beginning of my dissertation. I hope to have shown that there are good reasons to suppose that egocentric spatial representations are similar to self-conscious states in that they share this distinctive, functional role non-contingently. What remains unclear up to now, however, is what more is needed for self-representational states to become self-conscious, i.e. to reflect an awareness by the subject about the self-representational character of these states itself. Having explicated a possible root of self-consciousness the task now would be to spell out the way and the steps that are necessary to get from the root to that what is rooted, i.e. from the self-representational egocentric space to mature self-consciousness, and this, of course, not in a developmental sense but in a conceptual sense.

References

Adolph, K.E. (1995): Psychophysical assessment of toddlers' ability to cope with slopes. In *Journal of Experimental Psychology: Human Perception & Performance* 21 (4), pp. 734–750.

Adolph, K.E.; Eppler, M.A.; Gibson, E.J. (1993): Crawling versus walking infants' perception of affordances for locomotion over sloping surfaces. In *Child Development* 64 (4), pp. 1158–1174.

Adolph, K.E.; Eppler, M.A.; Marin, L. et al. (2000): Exploration in the service of prospective control. In *Infant Behavior and Development* 23 (3–4), pp. 441–460.

Aglioti, S.; DeSouza, J.F. X.; Goodale, M.A. (1995): Size-contrast illusions deceive the eye but not the hand. In *Current Biology* 5 (6), pp. 679–685.

Anderson, M.L. and Perlis, D.R. (2005): The roots of self-awareness. In *Phenomenology and the Cognitive Sciences* 4 (3), pp. 297–333.

Auvray, M.; Hanneton, S.; Lenay, C. et al. (2005): There is something out there. Distal attribution in sensory substitution, twenty years later. In *Journal of Integrative Neuroscience* 4 (4), pp. 505–521.

Auvray, M.; Hanneton, S.; O'Regan, J.K. (2007): Learning to perceive with a visuo-auditory substitution system: localisation and object recognition with 'the vOICe'. In *Perception* 36 (3), pp. 416–430.

Bach-y-Rita, P. (1972): *Brain Mechanisms in Sensory Substitution.* New York: Academic Press.

Bach-y-Rita, P. (1983): Tactile vision substitution: Past and future. In *International Journal of Neuroscience* 19 (1–4), pp. 29–36.

Bach-y-Rita, P. (1984): The relationship between motor processes and cognition in tactile vision substitution. In W. Prinz and D.A. Allport (Eds.): *Cognition and motor processes.* Berlin, Heidelberg: Springer, pp. 149–160.

Bach-y-Rita, P. (2004): Tactile sensory substitution studies. In *Annals of the New York Academy of Sciences* 1013 (1), pp. 83–91.

Bach-y-Rita, P. and Kercel, S.W. (2002): Sensori-'motor' coupling by observed and imagined movement. In *Intellectica* 35, pp. 287–297.

Bach-y-Rita, P. and Kercel, S.W. (2003): Sensory substitution and the human-machine interface. In *Trends in Cognitive Sciences* 7 (12), pp. 541–546.

Bach-y-Rita, P.; Tyler, M.E.; Kaczmarek, K.A. (2003): Seeing with the brain. In *International Journal of Human-Computer Interaction* 15 (2), pp. 285–295.

Bain, A. (1855): *The Aenses and the Intellect.* London: John W. Parker and Son West Strand.

Baldwin, T. (1995): Objectivity, causality and agency. In J.L. Bermúdez, A.J. Marcel, N. Eilan (Eds.): *The Body and the Self.* Cambridge, Mass.: MIT Press, pp. 107–126.

Baldwin, T. (2003): Perception and agency. In J. Roessler and N. Eilan (Eds.): *Agency and Self-Awareness. Issues in Philosophy and Psychology.* Oxford: Clarendon Press, pp. 188–200.

Baldwin, T. and Bell, D. (1988): Phenomenology, solipsism and egocentric thought. In *Proceedings of the Aristotelian Society, Supplementary Volumes* 62, pp. 27–60.

Beckermann, A. (2003): Self-Consciousness in cognitive systems. In C. Kanzian (Ed.): *Persons. An Interdisciplinary Approach.* Proceedings of the 25th International Wittgenstein Symposium 11th to 17th August Kirchberg am Wechsel (Austria). Kirchberg am Wechsel: Österreichische Ludwig-Wittgenstein-Gesellschaft, pp. 175–188.

Békésy, G. von (1964): Olfactory analogue to directional hearing. In *Journal of applied physiology* 19, pp. 369–373.

Belin, P. and Zatorre, R.J. (2000): 'What', 'where' and 'how' in auditory cortex. In *Nature Neuroscience* 3, pp. 965–966.

Bennett, K. (2011): Construction area (no hard hat required). In *Philosophical Studies* 154 (1), pp. 79–104.

Bermúdez, J.L. (1995): Transcendental arguments and psychology. The example of O'Shaughnessy on intentional action. In *Metaphilosophy* 26 (4), pp. 379–401.

Bermúdez, J.L. (1998): *The Paradox of Self-Consciousness.* Cambridge, Mass.: MIT Press.

Bermúdez, J.L. (2003): The phenomenology of bodily perception. In *Theoria et Historia Scientiarum: International Journal for Interdisciplinary Studies* 7 (1), pp. 43–52.

Bermúdez, J.L. (2011): Bodily awareness and self-consciousness. In S. Gallagher (Ed.): *The Oxford Handbook of the Self.* Oxford: Oxford University Press, pp. 157–180.

Berti, A. and Frassinetti, F. (2000): When far becomes near: Remapping of space by tool use. In *Journal of Cognitive Neuroscience* 12 (3), pp. 415–420.

Bhalla, M. and Proffitt, D.R. (1999): Visual-motor recalibration in geographical slant perception. In *Journal of Experimental Psychology: Human Perception & Performance* 25 (4), pp. 1076–1096.

Bower, T.G. R. (1974): *Development in Infancy.* San Francisco: Freeman.

Briscoe, R. (2009): Egocentric spatial representation in action and perception. In *Philosophy and Phenomenological Research* 79 (2), pp. 423–460.

Briscoe, R. (2011): Mental imagery and the varieties of amodal perception. In *Pacific Philosophical Quarterly* 92 (2), pp. 153–173.

Briscoe, R. (forthcoming): Bodily action and distal attribution in sensory substitution. In F. Macpherson (Ed.): *Sensory Substitution and Augmentation*. Oxford: Oxford University Press (Proceedings of the British Academy).

Brogaard, B. (2011): Conscious vision for action versus unconscious vision for action? In *Cognitive Science* 35 (6), pp. 1076–1104.

Brown, D.; Macpherson, T.; Ward, J. (2011): Seeing with sound? Exploring different characteristics of a visual-to-auditory sensory substitution device. In *Perception* 40 (9), pp. 1120–1135.

Burge, T. (2010): *Origins of Objectivity*. Oxford: Oxford University Press.

Cairns, D. (1976): *Conversations with Husserl and Fink*. The Hague: Nijhoff (Phaenomenologica, 66).

Campbell, J. (1993): The role of physical objects in spatial thinking. In N. Eilan, R.A. McCarthy, B. Brewer (Eds.): *Spatial Representation. Problems in Philosophy and Psychology*. Oxford, Cambridge, Mass.: Blackwell, pp. 65–95.

Campbell, J. (1994): *Past, Space, and Self*. Cambridge, Mass.: MIT Press.

Campbell, J. (2002): *Reference and Consciousness*. Oxford, New York: Oxford University Press.

Campbell, J. (2003): The role of demonstratives in action-explanation. In J. Roessler and N. Eilan (Eds.): *Agency and Self-awareness. Issues in Philosophy and Psychology*. Oxford: Clarendon Press, pp. 150–164.

Capelle, C.; Trullemans, C.; Arno, P. et al. (1998): A real-time experimental prototype for enhancement of vision rehabilitation using auditory substitution. In *IEEE transactions on bio-medical engineering* 45 (10), pp. 1279–1293.

Carman, T. (1999): The body in Husserl and Merleau-Ponty. In *Philosophical Topics* 27, pp. 205–226.

Casla, M.; Blanco, F.; Travieso, D. (1999): Haptic perception of geometric illusions by persons who are totally congenitally blind. In *Journal of Visual Impairment & Blindness* 93 (9), pp. 583–588.

Cassam, Q. (1999): *Self and World*. Oxford: Clarendon Press.

Casullo, A. (1986): The spatial structure of perceptual space. In *Philosophy and Phenomenological Research* 46 (4), pp. 665–671.

Christiansen, J.H.; Christensen, J.; Grünbaum, T. et al. (2014): A common representation of spatial features drives action and perception: grasping and judging object features within trials. In *PLoS ONE* 9 (5), pp. e94744.

Claesges, U. (1964): *Edmund Husserls Theorie der Raumkonstitution*. Den Haag: Nijhoff (Phaenomenologica, 19).

Clark, A. (1989): *Microcognition. Philosophy, Cognitive Science and Parallel Distributed Processing*. Cambridge, Mass.: MIT Press.

Clark, A. (2001): Visual experience and motor action: Are the bonds too tight? In *Philosophical Review* 110 (4), pp. 495–519.
Cole, J. and Paillard, J. (1995): Living without touch and peripheral information about body position and movement. Studies with deafferented subjects. In J.L. Bermúdez, A.J. Marcel, N. Eilan (Eds.): *The Body and the Self*. Cambridge, Mass.: MIT Press, pp. 245–266.
Colombo, M. (2013): Constitutive relevance and the personal/subpersonal distinction. In *Philosophical Psychology* 26 (4), pp. 547–570.
Conrad, K. (1933): Das Körperschema. In *Zeitschrift für die gesamte Neurologie und Psychiatrie* 147 (1), pp. 346–369.
Corlett, J.T.; Byblow, W.; Taylor, B. (1990): The effect of perceived locomotor constraints on distance estimation. In *Journal of Motor Behavior* 22 (3), pp. 347–360.
Cummins, R. (1984): *The Nature of Psychological Explanation*. 2nd ed. Cambridge, Mass.: MIT Press.
Cummins, R. (2000): "How does it work?" versus "what are the laws?". Two conceptions of psychological explanation. In R.A. Wilson and F.C. Keil (Eds.): *Explanation and Cognition*. Cambridge, Mass., London: MIT Press, pp. 117–144.
Davies, M. (2010): Double dissociation: Understanding its role in cognitive neuropsychology. In *Mind & Language* 25 (5), pp. 500–540.
Degenaar, J. (2014): Through the inverting glass: first-person observations on spatial vision and imagery. In *Phenomenology and the Cognitive Sciences* 13 (2), pp. 373–393.
Dennett, D.C. (1969): *Content and Consciousness*. London: Routledge & Kegan Paul.
Dijkerman, H.C. and de Haan, E.H. F. (2007): Somatosensory processes subserving perception and action. In *The Behavioral and brain sciences* 30 (2), pp. 189–201.
Dokic, J. (2013): The role of noetic feelings in sensory substitution. Sensory Substitution and Augmentation Conference. The British Academy. London, 3/27/2013. Available online at http://www.gla.ac.uk/schools/humanities/research/philosophyresearch/cspe/events/sensorysubstitutionandaugmentationconference/, checked on 8/18/2014.
Dokic, J. (2015): Perception and space. In M. Matthen (Ed.): *Oxford Handbook for the Philosophy of Perception*. Oxford:: Oxford University Press.
Dolezal, H. (1982): *Living in a World Transformed. Perceptual and Performatory Adaption to Visual Distortion*. New York: Academic Press.
Dretske, F.I. (1995): *Naturalizing the Mind*. Cambridge, Mass.: MIT Press.
Drummond, J.J. (1979): On seeing *a* material thing *in* space: The role of kinaesthesis in visual perception. In *Philosophy and Phenomenological Research* 40 (1), pp. 19–32.

Durgin, F.H.; Baird, J.A.; Greenburg, M. et al. (2009): Who is being deceived? The experimental demands of wearing a backpack. In *Psychonomic Bulletin & Review* 16 (5), pp. 964–969.

Durgin, F.H.; Klein, B.; Spiegel, A. et al. (2012): The social psychology of perception experiments: Hills, backpacks, glucose, and the problem of generalizability. In *Journal of Experimental Psychology: Human Perception & Performance* 38 (6), pp. 1582–1595.

Epstein, W.; Hughes, B.; Schneider, S. et al. (1986): Is there anything out there? A study of distal attribution in response to vibrotactile stimulation. In *Perception* 15 (3), pp. 275–284.

Ernst, M.O. and Bülthoff, H.H. (2004): Merging the senses into a robust percept. In *Trends in Cognitive Sciences* 8 (4), pp. 162–169.

Evans, G. (1982): *The Varieties of Reference.* Edited by J.H. McDowell. Oxford: Clarendon Press.

Evans, G. (1985): Molyneux's question. In Gareth Evans: *Collected Papers.* Edited by A. Phillips. Oxford: Clarendon Press, pp. 364–399.

Farina, M. (2013): Neither touch nor vision: sensory substitution as artificial synaesthesia? In *Biology & Philosophy* 28 (4), pp. 639–655.

Farnè, A. and Làdavas, E. (2000): Dynamic size-change of hand peripersonal space following tool use. In *NeuroReport* 11 (8), pp. 1645–1649.

Ferencz-Flatz, C. (2014): Husserls Begriff der Kinästhese und seine Entwicklung. In *Husserl Studies* 30 (1), pp. 21–45.

Fichte, J.G. (1970): *Versuch einer neuen Darstellung der Wissenschaftslehre.* In Johann Gottlieb Fichte: Gesamtausgabe der Bayerischen Akademie der Wissenschaften. 1. Werke. Werke 1797–1798. Edited by R. Lauth. Stuttgart-Bad Cannstatt: Frommann.

Firestone, C. (2013): How "paternalistic" is spatial perception? Why wearing a heavy backpack doesn't – and couldn't – make Hills look steeper. In *Perspectives on Psychological Science* 8 (4), pp. 455–473.

Firestone, C. and Scholl, B.J. (2014): "Top-down" effects where none should be found. The El Greco Fallacy in perception research. In *Psychological Science* 25 (1), pp. 38–46.

Franz, V.H. and Gegenfurtner, K.R. (2008): Grasping visual illusions: Consistent data and no dissociation. In *Cognitive Neuropsychology* 25 (7–8), pp. 920–950.

Gallagher, S. (1995): Body schema and intentionality. In J.L. Bermúdez, A.J. Marcel, N. Eilan (Eds.): *The Body and the Self.* Cambridge, Mass.: MIT Press, pp. 225–244.

Gallagher, S. (2003): Bodily self-awareness and object perception. In *Theoria et Historia Scientiarum: International Journal for Interdisciplinary Studies* 7, pp. 53–68.

Gallagher, S. (2005): *How the Body Shapes the Mind.* Oxford: Clarendon Press.

Gallagher, S. (2009): The intrinsic spatial frame of reference. In H.L. Dreyfus (Ed.): *A Companion to Phenomenology and Existentialism.* Oxford: Wiley-Blackwell, pp. 346–355.

Gallistel, C.R. (1980): *The Organization of Action. A New Synthesis.* Hillsdale, N.J., New York: L. Erlbaum Associates.

Gendler, T. (2013): "Imagination". *The Stanford Encyclopedia of Philosophy* (Fall 2013 Edition). Edited by E.N. Zalta. Available online at http://plato.stanford.edu/archives/fall2013/entries/imagination/, checked on 9/19/2014.

Gentilucci, M.; Chieffi, S.; Daprati, E. et al. (1996): Visual illusion and action. In *Neuropsychologia* 34 (5), pp. 369–376.

Gibson, J.J. (1950): *The Perception of the Visual World.* Boston: Mifflin.

Gibson, J.J. (1979): *The Ecological Approach to Visual Perception.* Boston: Houghton Mifflin.

Glover, S. (2004): Separate visual representations in the planning and control of action. In *Behavioral and Brain Sciences* 27 (1), pp. 3–24.

Goodale, M.A. (2011): Transforming vision into action. Vision Research 50th Anniversary Issue: Part 2. In *Vision Research* 51 (13), pp. 1567–1587.

Goodale, M.A. and Haffenden, A. (1998): Frames of reference for perception and action in the human visual system. In *Neuroscience & Biobehavioral Reviews* 22 (2), pp. 161–172.

Goodale, M.A. and Humphrey, G.K. (1998): The objects of action and perception. In *Cognition* 67 (1–2), pp. 181–207.

Goodale, M.A. and Milner, A.D. (1992): Separate visual pathways for perception and action. In *Trends in Neurosciences* 15 (1), pp. 20–25.

Goodale, M.A.; Milner, A.D.; Jakobson, L.S. et al. (1991): A neurological dissociation between perceiving objects and grasping them. In *Nature* 349 (6305), pp. 154–156.

Grave, D.D. J. de; Brenner, E.; Smeets, J.B. et al. (2011): Using a stick does not necessarily alter judged distances or reachability. In *PLoS ONE* 6 (2), pp. e16697.

Grush, R. (2000): Self, world and space: The meaning and mechanisms of ego- and allocentric spatial representation. In *Brain and Mind* 1, pp. 59–92.

Grush, R. (2007): Skill theory v2.0: dispositions, emulation, and spatial perception. In *Synthese* 159, pp. 389–416.

Guarniero, G. (1974): Experience of tactile vision. In *Perception* 3 (1), pp. 101–104.

Head, H. and Holmes, G. (1911): Sensory disturbances form cerebral lesions. In *Brain* 34 (2–3), pp. 102–254.

Heidegger, M. (1972): *Sein und Zeit.* 12., unveränd. Aufl. Tübingen: Niemeyer.

Heidegger, M. (1979): Prolegomena zur Geschichte des Zeitbegriffs. In Martin Heidegger: *Gesamtausgabe*, II. Abteilung: *Vorlesungen 1923–1944*, Bd. 20. Edited by P. Jaeger. Frankfurt am Main: Klostermann.

Heidegger, M. (2010): *Being and Time*. A revised edition of the Stambaugh translation. Translated by J. Stambaugh; revised and with a forword by D. J. Schmidt. Albany: State University of New York Press.

Heil, J. (1987): The Molyneux question. In *Journal for the Theory of Social Behaviour* 17 (3), pp. 227–241.

Held, R. (2009): Visual-haptic mapping and the origin of cross-modal identity. In *Optometry and vision science: official publication of the American Academy of Optometry* 86 (6), pp. 595–598.

Hofmann, F. (2001): The reference of de re representations. In *Grazer Philosophische Studien* 62 (1), pp. 83–101.

Hofmann, F. and Pöhlmann, F. (2013): Seeing oneself through the eyes of others. Beckermann on self-consciousness. In *Philosophia Naturalis* 50 (1), pp. 25–43.

Hornsby, J. (1980): *Actions*. London, Boston: Routledge & Kegan Paul.

Hurley, S. and Noë, A. (2003): Neural plasticity and consciousness. In *Biology & Philosophy* 18 (1), pp. 131–168.

Hurley, S.L. (1998): *Consciousness in Action*. Cambridge Mass.: Harvard University Press.

Hurley, S.L. (2001): Perception and action: alternative views. In *Synthese* 129 (1), pp. 3–40.

Husserl, E. (1940a): Notizen zur Raumkonstitution. In *Philosophy and Phenomenological Research* 1 (1), pp. 21–37.

Husserl, E. (1940b): Notizen zur Raumkonstitution. Continued. In *Philosophy and Phenomenological Research* 1 (2), pp. 217–226.

Husserl, E. (1954): *Die Krisis der europäischen Wissenschaften und die transzendentale Phänomenologie*. In Husserliana / Edmund Husserl. Gesammelte Werke, VI. Edited by W. Biemel. Den Haag: Nijhoff.

Husserl, E. (1973): Zur Phänomenologie der Intersubjektivität. Texte aus d. Nachlaß ; 2. Teil: 1921–1928. In Husserliana / Edmund Husserl. Gesammelte Werke, 14. Edited by H.L. van Breda. Den Haag: Nijhoff.

Husserl, E. (1983): Philosophische Versuche über den Raum. In Edmund Husserl: *Studien zur Arithmetik und Geometrie. Texte aus d. Nachlaß (1886–1901)*. Husserliana / Edmund Husserl. Gesammelte Werke, XXI. Edited by I. Strohmeyer. Dordrecht: Nijhoff, pp. 261–310.

Husserl, E. (1991): *Ding und Raum. Vorlesungen 1907*. Edited by K.-H. Hahnengress, S. Rapic. Hamburg: F. Meiner.

Husserl, E. (2008a): Beilage XXX. Zur haptischen Konstitution der praktischen Welt. Der Vorzug der Tastwahrnehmungen vor den visuellen Wahrnehmungen. In Edmund Husserl: *Die Lebenswelt. Auslegungen der vorgegebenen Welt und ihrer Konstitution. Texte aus dem Nachlass (1916–1937).* Husserliana / Edmund Husserl. Gesammelte Werke, XXXIX. Edited by R. Sowa. Dordrecht: Springer, pp. 396–400.

Husserl, E. (2008b): *Die Lebenswelt. Auslegungen der vorgegebenen Welt und ihrer Konstitution. Texte aus dem Nachlass (1916–1937).* Husserliana / Edmund Husserl. Gesammelte Werke, XXXIX. Edited by R. Sowa. Dordrecht: Springer.

Hutchison, J.J. and Loomis, J.M. (2006): Does energy expenditure affect the perception of egocentric distance? A failure to replicate experiment 1 of Proffitt, Stefanucci, Banton, and Epstein (2003). In *The Spanish Journal of Psychology* 9 (2), pp. 332–339.

Iriki, A.; Tanaka, M.; Iwamura, Y. (1996): Coding of modified body schema during tool use by macaque postcentral neurones. In *NeuroReport* 7 (14), pp. 2325–2330.

Jackson, G.B. (2014): Skillful action in peripersonal space. In *Phenomenology and the Cognitive Sciences* 13 (2), pp. 313–334.

Jeannerod, M. (1997): *The Cognitive Neuroscience of Action.* Oxford, Cambridge, Mass.: Blackwell.

Jeannerod, M. (2001): Neural simulation of action: A unifying mechanism for motor cognition. In *NeuroImage* 14 (1), pp. S103–S109.

Kant, I. (1998): *Critique of Pure Reason.* Edited by P. Guyer, A.W. Wood. Cambridge, New York: Cambridge University Press.

Karnath, H.-O. and Thier, P. (Eds.) (2012): *Kognitive Neurowissenschaften.* 3rd ed. Berlin, Heidelberg: Springer.

Kirsch, W. and Kunde, W. (2013): Visual near space is scaled to parameters of current action plans. In *Journal of Experimental Psychology: Human Perception & Performance* 39 (5), pp. 1313–1325.

Klein, C. (2007): Kicking the Kohler habit. In *Philosophical Psychology* 20 (5), pp. 609–619.

Kohler, I. (1951): *Über Aufbau und Wandlungen der Wahrnehmungswelt. Insbesondere über „bedingte Empfindungen“.* Sitzungsberichte / Österreichische Akademie der Wissenschaften, Philosophisch-Historische Klasse, 2271. Wien: Rohrer.

Kohler, I. (1964): *The Formation and Transformation of the Perceptual World.* New York: International Universities Press.

Króliczak, G.; Heard, P.; Goodale, M.A. et al. (2006): Dissociation of perception and action unmasked by the hollow-face illusion. In *Brain Research* 1080 (1), pp. 9–16.

Kruse, L. (1974): Räumliche Umwelt. Die Phänomenologie des räumlichen Verhaltens als Beitrag zu einer psychologischen Umwelttheorie. Berlin, New York: W. de Gruyter.

Làdavas, E. and Serino, A. (2008): Action-dependent plasticity in peripersonal space representations. In *Cognitive Neuropsychology* 25 (7–8), pp. 1099–1113.

Lailach-Hennrich, A. (2013): Kein Selbst ohne die Anderen? Was wir immer noch von G. H. Meads Theorie des Selbst lernen können. In *Zeitschrift für philosophische Forschung* 67 (1), pp. 61–88.

Lang, S. (2010): *Spontaneitat des Selbst.* Göttingen: V & R Unipress.

Levy-Tzedek, S.; Hanassy, S.; Abboud, S. et al. (2012): Fast, accurate reaching movements with a visual-to-auditory sensory substitution device. In *Restorative Neurology and Neuroscience* 30, pp. 313–323.

Lewin, K. (1982): Vektoren, kognitive Prozesse und Mr. Tolmans Kritik. In Kurt Lewin: *Feldtheorie.* Kurt-Lewin-Werkausgabe, 4. Edited by C.-F. Graumann. Bern: Huber, pp. 99–131.

Linden, D.E. J.; Kallenbach, U.; Heinecke, A. et al. (1999): The myth of upright vision. A psychophysical and functional imaging study of adaptation to inverting spectacles. In *Perception* 28 (4), pp. 469–481.

Linkenauger, S.A.; Witt, J.K.; Stefanucci, J. et al. (2006): Ease to grasp an object affects perceived distance. In *Journal of Vision* 6 (6), pp. 724a.

Longo, M.R. and Lourenco, S.F. (2006): On the nature of near space: Effects of tool use and the transition to far space. In *Neuropsychologia* 44 (6), pp. 977–981.

Mach, E. (1922): Die Analyse der Empfindungen und das Verhältnis des Physischen zum Psychischen. 9th ed. Jena: Verlag Gustav Fischer.

Macpherson, F. (2013): Introduction to the issues. Sensory Substitution and Augmentation Conference. The British Academy. London, 3/26/2013. Available online at http://www.gla.ac.uk/schools/humanities/research/philosophyresearch/cspe/events/sensorysubstitutionandaugmentationconference/, checked on 8/13/2014.

Majid, A.; Bowerman, M.; Kita, S. et al. (2004): Can language restructure cognition? The case for space. In *Trends in Cognitive Sciences* 8 (3), pp. 108–114.

Mallot, H.A. and Basten, K. (2009): Embodied spatial cognition. Biological and artificial systems. In *Image and Vision Computing* 27, pp. 1658–1670.

Malpas, J.E. (1999): *Place and Experience. A Philosophical Topography.* New York: Cambridge University Press.

Martin, J.-R. and Le Corre, F. (2015): Sensory substitution is substitution. In *Mind & Language* 30 (2), pp. 209–233.

Mattens, F. (2010): Kinästhese. In H.-H. Gander (Ed.): *Husserl-Lexikon*. Darmstadt: Wissenschaftliche Buchgesellschaft, pp. 170–172.

Matthen, M. (2005): *Seeing, Doing, and Knowing. A Philosophical Theory of Sense Perception*. Oxford: Oxford University Press.

Matthen, M.; Macpherson, F.; Smith, B. et al. (2013): Round table discussion. Sensory Substitution and Augmentation Conference. The British Academy. London, 3/28/2013. Available online at http://www.gla.ac.uk/schools/humanities/research/philosophyresearch/cspe/events/sensorysubstitutionandaugmentationconference/, checked on 8/13/2014.

Meijer, P.B. (1992): An experimental system for auditory image representations. In *IEEE transactions on bio-medical engineering* 39 (2), pp. 112–121.

Merleau-Ponty, M. (1945): *Phénoménologie de la perception*. Paris: Gallimard.

Merleau-Ponty, M. (1974): *Phänomenologie der Wahrnehmung*. Photomech. Nachdr. Edited by R. Boehm. Berlin: De Gruyter.

Merleau-Ponty, M. (2002): *Phenomenology of Perception*. Translated by C. Smith. London, New York: Routledge.

Milner, A.D.; Dijkerman, H.C.; Pisella, L. et al. (2001): Grasping the past: delay can improve visuomotor performance. In *Current Biology* 11 (23), pp. 1896–1901.

Milner, A.D. and Goodale, M.A. (1995): *The Visual Brain in Action*. Oxford, New York: Oxford University Press.

Milner, A.D. and Goodale, M.A. (2006): *The Visual Brain in Action*. 2nd ed. Oxford, New York: Oxford University Press.

Milner, A.D. and Goodale, M.A. (2008): Two visual systems re-viewed. In *Neuropsychologia* 46 (3), pp. 774–785.

Mossio, M. and Taraborelli, D. (2008): Action-dependent perceptual invariants: From ecological to sensorimotor approaches. In *Consciousness and Cognition* 17 (4), pp. 1324–1340.

Musholt, K. (2012): Self-consciousness and intersubjectivity. In *Grazer Philosophische Studien* 84, pp. 75–101.

Nanay, B. (2015): Trompe l'oeil and the dorsal/ventral account of picture perception. In *Review of Philosophy and Psychology* 6 (1), pp. 181–197.

Neisser, U. (1988): Five kinds of self-knowledge. In *Philosophical Psychology* 1 (1), pp. 35–59.

Neisser, U. (1995): Criteria for an ecological self. In P. Rochat (Ed.): *The Self in Infancy. Theory and Research*. Amsterdam, New York: Elsevier, pp. 17–34.

Neumann, G. (1999): *Die phänomenologische Frage nach dem Ursprung der mathematisch-naturwissenschaftlichen Raumauffassung bei Husserl und Heidegger*. Berlin: Duncker & Humblot.

Noë, A. (2004): *Action in Perception.* Cambridge, Mass.: MIT Press.

Noë, A. (2008): Reply to Campbell, Martin, and Kelly. In *Philosophy and Phenomenological Research* 76 (3), pp. 691–706.

Nozick, R. (1981): *Philosophical Explanations.* Oxford: Clarendon.

O'Brien, L. (2007): *Self-Knowing Agents.* Oxford: Oxford University Press.

O'Regan, J.K. and Noë, A. (2001): A sensorimotor account of vision and visual consciousness. In *Behavioral and Brain Sciences* 24 (5), pp. 939–973.

Orne, M.T. (1962): On the social psychology of the psychological experiment: With particular reference to demand characteristics and their implications. In *American Psychologist* 17 (11), pp. 776–783.

Orne, M.T. (2009): Demand Characteristics and the concept of quasi-controls. In R. Rosenthal and R.L. Rosnow (Eds.): *Artifacts in Behavioral Research.* Robert Rosenthal and Ralph L. Rosnow's classic books. New York: Oxford University Press, pp. 110–137.

O'Shaughnessy, B. (1992): The diversity and unity of action and perception. In T. Crane (Ed.): *The Contents of Experience. Essays on Perception.* Cambridge, UK, New York: Cambridge University Press, pp. 216–266.

O'Shaughnessy, B. (1995): Proprioception and the body image. In J.L. Bermúdez, A.J. Marcel, N. Eilan (Eds.): *The Body and the Self.* Cambridge, Mass.: MIT Press, pp. 175–203.

O'Shaughnessy, B. (2008): *The Will. A Dual Aspect Theory.* 2nd ed. Cambridge u.a: Cambridge University Press.

Osiurak, F.; Morgado, N.; Palluel-Germain, R. (2012): Tool use and perceived distance: when unreachable becomes spontaneously reachable. In *Experimental Brain Research* 218 (2), pp. 331–339.

Peacocke, C. (1983): *Sense and Content. Experience, Thought, and their Relations.* Oxford, New York: Clarendon Press; Oxford University Press.

Peacocke, C. (2014): The Mirror of the World. Subjects, Consciousness, and Self-Consciousness. Oxford: Oxford University Press.

Perry, J. (1986): Thought without representation. In *Proceedings of the Aristotelian Society* 137, pp. 137–152.

Perry, J. (1993a): Postscript to "Thought without representation". In J. Perry (Ed.): *The Problem of the Essential Indexical. And other Essays.* New York: Oxford University Press, pp. 219–225.

Perry, J. (1993b): The problem of the essential indexical. In J. Perry (Ed.): *The Problem of the Essential Indexical. And other Essays.* New York: Oxford University Press, pp. 33–52.

Perry, J. (1998a): Indexicals, contexts and unarticulated constituents. In *Proceedings of the 1995 CSLI-Amsterdam Logic, Language and Computation Conference*. CSLI Publications.

Perry, J. (1998b): Myself and I. In M. Stamm (Ed.): *Philosophie in synthetischer Absicht. Synthesis in Mind*. Stuttgart: Klett-Cotta, pp. 83–103.

Perry, J. (2011): On Knowing one's Self. In S. Gallagher (Ed.): *The Oxford Handbook of the Self*. Oxford: Oxford University Press, pp. 372–393.

Pick, A. (1922): Störung der Orientierung am eigenen Körper. In *Psychologische Forschung* 1 (1), pp. 303–318.

Pick Jr, H.L. (1993): Organization of spatial knowledge in children. In N. Eilan, R.A. McCarthy, B. Brewer (Eds.): *Spatial representation. Problems in Philosophy and Psychology*. Oxford, Cambridge, Mass.: Blackwell, pp. 31–42.

Piryankova, I.V.; Wong, H.Y.; Linkenauger, S.A. et al. (2014): Owning an overweight or underweight body: distinguishing the physical, experienced and virtual body. In *PLoS ONE* 9 (8), pp. e103428.

Pisella, L.; Gréa, H.; Tilikete, C. et al. (2000): An 'automatic pilot' for the hand in human posterior parietal cortex: toward reinterpreting optic ataxia. In *Nature Neuroscience* 3 (7), pp. 729–736.

Pitcher, G. (1971): *A Theory of Perception*. Princeton, N. J.: Princeton University Press.

Place, U.T. (1996): Intentionality as the mark of the dispositional. In *Dialectica* 50 (2), pp. 91–120.

Poincaré, H. (2010): *Science and Method*. Cosimo Classics.

Porter, J.; Anand, T.; Johnson, B. et al. (2005): Brain mechanisms for extracting spatial information from smell. In *Neuron* 47 (4), pp. 581–592.

Proffitt, D.R. (2006): Embodied perception and the economy of action. In *Perspectives on Psychological Science* 1 (2), pp. 110–122.

Proffitt, D.R. (2008): An action-specific approach to spatial perception. In R.L. Klatzky, B. MacWhinney, M. Behrmann (Eds.): *Embodiment, Ego-Space, and Action*. New York: Psychology Press, pp. 179–202.

Proffitt, D.R. and Linkenauger, S.A. (2013): Perception Viewed as a Phenotypic Expression. In W. Prinz, M. Beisert, A. Herwig (Eds.): *Action Science*. Cambridge: MIT Press, pp. 171–197.

Proffitt, D.R.; Stefanucci, J.; Banton, T. et al. (2003): The Role of effort in perceiving distance. In *Psychological Science* 14 (2), pp. 106–112.

Proulx, M.J.; Stoerig, P.; Ludowig, E. et al. (2008): Seeing 'where' through the ears: Effects of learning-by-doing and long-term sensory deprivation on localization based on image-to-sound substitution. In *PLoS ONE* 3 (3).

Rauschecker, J.P. (1998): Cortical processing of complex sounds. In *Current opinion in neurobiology* 8 (4), pp. 516–521.

Renier, L.; Bruyer, R.; Volder, A.G. de (2006): Vertical-horizontal illusion present for sighted but not early blind humans using auditory substitution of vision. In *Perception & Psychophysics* 68 (4), pp. 535–542.

Renier, L.; Laloyaux, C.; Collignon, O. et al. (2005): The Ponzo illusion with auditory substitution of vision in sighted and early-blind subjects. In *Perception* 34 (7), pp. 857–867.

Renier, L. and Volder, A.G. de (2005): Cognitive and brain mechanisms in sensory substitution of vision. A contribution to the study of human perception. In *Journal of Integrative Neuroscience* 4 (4), pp. 489–503.

Renier, L. and Volder, A.G. de (2010): Vision substitution and depth perception. Early blind subjects experience visual perspective through their ears. In *Disability and Rehabilitation: Assistive Technology* 5 (3), pp. 175–183.

Richards, W.M. (1984): Self-consciousness and agency. In *Synthese* 61 (2), pp. 149–171.

Rieser, J.J.; Pick, H.L.; Ashmead, D.H. et al. (1995): Calibration of human locomotion and models of perceptual-motor organization. In *Journal of Experimental Psychology: Human Perception & Performance* 21 (3), pp. 480–497.

Rizzolatti, G.; Fadiga, L.; Fogassi, L. et al. (1997): The space around us. In *Science* 277 (5323), pp. 190–191.

Rochat, P. (1995): Early objectification of the self. In P. Rochat (Ed.): *The Self in Infancy. Theory and Research.* Amsterdam, New York: Elsevier, pp. 53–71.

Rochat, P. (2004): *The Infant's World.* Cambridge, Mass.: Harvard University Press.

Rochat, P. (2009): *Others in Mind. Social Origins of Self-Consciousness.* Cambridge, New York: Cambridge University Press.

Rochat, P.; Goubet, N.; Senders, S.J. (1999): To reach or not to reach? Perception of body effectivities by young infants. In *Infant and Child Development* 8 (3), pp. 129–148.

Rock, I. (1985): *The Logic of Perception.* Cambridge, Mass.: MIT Press.

Rossetti, Y.; Hisaaki, O.; Blangero, A. et al. (2010): Why does the perception-action functional dichotomy not match the ventral-dorsal streams anatomical segregation: optic ataxia and the function of the dorsal stream. In N. Gangopadhyay, M. Madary, F. Spicer (Eds.): *Perception, Action, and Consciousness. Sensorimotor Dynamics and Two Visual Systems.* New York: Oxford University Press, pp. 163–181.

Russell, R. and Durgin, F.H. (2008): Demand characteristics, not effort: The role of backpacks in judging distance. In *Journal of Vision* 8 (6), p. 755.

Sacks, O.W. (1986): *The Man who Mistook his Wife for a Hat.* London: Picador.

Scheler, M. (1928): Idealismus - Realismus. In *Philosophischer Anzeiger* 2, pp. 255–324.

Scheler, M. (1960): Erkenntnis und Arbeit. Eine Studie über Wert und Grenzen des pragmatischen Motivs in der Erkenntnis der Welt. In Max Scheler: *Die Wissensformen und die Gesellschaft.* 2nd ed. Max Scheler. Gesammelte Werke, 8. Edited by M. Scheler. Bern: Francke, pp. 191–382.

Scheler, M. (1973): Idealism and Realism. In Max Scheler: *Selected Philosophical Essays.* Translated, with an introd., by D.R. Lachterman. Evanston: Northwestern University Press, pp. 288–356.

Scheler, M. (1976): Die Stellung des Menschen im Kosmos. In Max Scheler: *Späte Schriften.* Max Scheler. Gesammelte Werke, 9. Edited by M.S. Frings. Bern: Francke, pp. 7–71.

Scheler, M. (2009): *The Human Place in the Cosmos.* Translated from the German by M.S. Frings. Introduction by E. Kelly. Evanston Ill: Northwestern University Press.

Schellenberg, S. (2007): Action and self-location in perception. In *Mind* 116 (463), pp. 603–632.

Schilder, P. (1923): *Das Körperschema. Ein Beitrag zur Lehre vom Bewusstsein des eigenen Körpers.* Berlin: Jul. Springer.

Schnall, S.; Harber, K.D.; Stefanucci, J.K. et al. (2008): Social support and the perception of geographical slant. In *Journal of Experimental Social Psychology* 44 (5), pp. 1246–1255.

Schnall, S.; Zadra, J.R.; Proffitt, D.R. (2010): Direct evidence for the economy of action: Glucose and the perception of geographical slant. In *Perception* 39 (4), pp. 464–482.

Searle, J.R. (1983): *Intentionality. An Essay in the Philosophy of Mind.* Cambridge, New York: Cambridge University Press.

Segond, H.; Weiss, D.; Sampaio, E. (2005): Human spatial navigation via a visuotactile sensory substitution system. In *Perception* 34 (10), pp. 1231–1249.

Sekiyama, K.; Hashimoto, K.; Sugita, Y. (2012): Visuo-somatosensory reorganization in perceptual adaptation to reversed vision. In *Acta Psychologica* 141 (2), pp. 231–242.

Sela, L. and Sobel, N. (2010): Human olfaction: a constant state of change-blindness. In *Experimental Brain Research* 205 (1), pp. 13–29.

Siegle, J.H. and Warren, W.H. (2010): Distal attribution and distance perception in sensory substitution. In *Perception* 39 (2), pp. 208–223.

Smith, J. (manuscript): Egocentric space.

Stefanucci, J.K.; Proffitt, D.R.; Clore, G.L. et al. (2008): Skating down a steeper slope: Fear influences the perception of geographical slant. In *Perception* 37 (2), pp. 321–323.

Stratton, G.M. (1897): Vision without inversion of the retinal image. In *Psychological Review* 4 (4, 5), pp. 341–360, 463–481.

Strawson, P.F. (1959): *Individuals. An Essay in Descriptive Metaphysics*. London: Methuen.

Stumpf, C. (1873): *Über den psychologischen Ursprung der Raumvorstellung*. Leipzig: Hirzel.

Taylor, C. (1978): The validity of transcendental arguments. In *Proceedings of the Aristotelian Society* 79, pp. 151–165.

Taylor, J.G. (1962): *The Behavioral Basis of Perception*. With a Mathematical Appendix to Chapter 8. New Haven: Yale University Press.

Tiemersma, D. (1982): 'Body-image' and 'body-schema' in the existential phenomenology of Merleau-Ponty. In *Journal of the British Society for Phenomenology* 13 (3), pp. 246–255.

Tugendhat, E. (1979): *Selbstbewusstsein und Selbstbestimmung. Sprachanalytische Interpretationen*. Frankfurt am Main: Suhrkamp.

Tugendhat, E. (2005): Über Selbstbewusstsein: Einige Missverständnisse. In T. Grundmann, F. Hofmann, C. Misselhorn, V.L. Waibel, V. Zanetti (Eds.): *Anatomie der Subjektivität. Bewusstsein, Selbstbewusstsein und Selbstgefühl*. Frankfurt am Main: Suhrkamp, pp. 247–254.

Uexküll, J. von (1920): *Theoretische Biologie*. Berlin: Paetel.

Uexküll, J. von (1973): *Theoretische Biologie*. Frankfurt a.M.: Suhrkamp.

Uexküll, J. von and Kriszat, G. (1956): *Streifzüge durch die Umwelten von Tieren und Menschen*. Hamburg: Rowohlt.

Ungerleider, L.G. and Mishkin, M. (1982): Object vision and spatial vision: Two cortical pathways. In D. Ingle, M.A. Goodale, R.J. W. Mansfield (Eds.): *Analysis of Visual Behavior*. Cambridge, Mass.: MIT Press, pp. 549–586.

van der Hoort, B.; Guterstam, A.; Ehrsson, H.H. et al. (2011): Being Barbie: The size of one's own body determines the perceived size of the wWorld. In *PLoS ONE* 6 (5), p. e20195.

Vignemont, F. de (2006): A review of Shaun Gallagher: *How the Body Shapes the Mind*. In *Psyche* 12 (1).

Vignemont, F. de (2010): Body schema and body image – Pros and cons. In *Neuropsychologia* 48 (3), pp. 669–680.

Vignemont, F. de (2014): A multimodal conception of bodily awareness. In *Mind* 123 (492), pp. 989–1020.

Wade, N.J. (2000): An upright man. In *Perception* 29 (3), pp. 253–257.

Ward, D.; Roberts, T.; Clark, A. (2011): Knowing what we can do: actions, intentions, and the construction of phenomenal experience. In *Synthese* 181 (3), pp. 375–394.

Ward, J. and Meijer, P.B. (2010): Visual experiences in the blind induced by an auditory sensory substitution device. In *Consciousness and Cognition* 19 (1), pp. 492–500.

Wertheimer, M. (1912): Experimentelle Studien über das Sehen von Bewegung. In *Zeitschrift für Psychologie* 60, pp. 321–378.

Witt, J.K. (2011): Action's effect on perception. In *Current Directions in Psychological Science* 20 (3), pp. 201–206.

Witt, J.K. and Dorsch, T.E. (2009): Kicking to bigger uprights: Field goal kicking performance influences perceived size. In *Perception* 38 (9), pp. 1328–1340.

Witt, J.K.; Linkenauger, S.A.; Bakdash, J.Z. et al. (2008): Putting to a bigger hole: golf performance relates to perceived size. In *Psychonomic Bulletin & Review* 15 (3), pp. 581–585.

Witt, J.K. and Proffitt, D.R. (2005): See the ball, hit the ball. Apparent ball size is correlated with batting average. In *Psychological Science* 16 (12), pp. 937–938.

Witt, J.K. and Proffitt, D.R. (2008): Action-specific influences on distance perception: A role for motor simulation. In *Journal of Experimental Psychology: Human Perception & Performance* 34 (6), pp. 1479–1492.

Witt, J.K.; Proffitt, D.R.; Epstein, W. (2004): Perceiving distance: A role of effort and intent. In *Perception* 33 (5), pp. 577–590.

Witt, J.K.; Proffitt, D.R.; Epstein, W. (2005): Tool use affects perceived distance, but only when you intend to use it. In *Journal of Experimental Psychology: Human Perception & Performance* 31 (5), pp. 880–888.

Wong, H.Y. (forthcoming): The body schema as a condition of possibility for action. In Y. Coello and M. Fischer (Eds.): *From Action to Concepts. Behavioral and Neural Foundations of Embodied Cognition*. Psychology Press.

Woods, A.J.; Philbeck, J.W.; Danoff, J.V. (2009): The various perceptions of distance: An alternative view of how effort affects distance judgments. In *Journal of Experimental Psychology: Human Perception & Performance* 35 (4), pp. 1104–1117.

Yablo, S. (1992): Cause and essence. In *Synthese* 93 (3), pp. 403–449.

Ylikoski, P. (2013): Causal and constitutive explanation compared. In *Erkenntnis* 78, pp. 277–297.

Index

FSC
www.fsc.org

MIX
Papier aus verantwortungsvollen Quellen
Paper from responsible sources
FSC® C105338

If you have any concerns about our products,
you can contact us on
ProductSafety@springernature.com

In case Publisher is established outside the EU,
the EU authorized representative is:
Springer Nature Customer Service Center GmbH
Europaplatz 3, 69115 Heidelberg, Germany

Printed by Libri Plureos GmbH
in Hamburg, Germany